KU-742-976

Timber: Properties and Uses

Timber: Properties and Uses

W.P.K. FINDLAY, D.Sc., F.I.W.Sc., F.I.Biol.

formerly Senior Principal Scientific
Officer at the Forest Products Research
Laboratory

CROSBY LOCKWOOD STAPLES
GRANADA PUBLISHING
London Toronto Sydney New York

Published by Granada Publishing Limited
and Crosby Lockwood Staples 1975
Reprinted 1978

Granada Publishing Limited
Frogmore, St Albans, Herts AL2 2NF
and
3 Upper James Street, London W1R 4BP
1221 Avenue of the Americas, New York, NY 10020 USA
117 York Street, Sydney, NSW 2000, Australia
100 Skyway Avenue, Toronto, Ontario, Canada M9W 3A6
Trio City, Coventry Street, Johannesburg 2001, South Africa
CML Centre, Queen & Wyndham Streets, Auckland 1, New Zealand

Copyright © 1975 W. P. K. Findlay

ISBN 0 258 97113 4 PB
0 258 96921 0 HB

Printed in Great Britain by
William Clowes & Sons Limited
London, Beccles and Colchester

All rights reserved. No part of this publication may be reproduced,
stored in a retrieval system, or transmitted, in any form or by any means,
electronic, mechanical, photocopying, recording or otherwise, without
the prior permission of the publishers.

Preface

When *Timber: Its Properties, Pests and Preservation* by F.Y. Henderson went out of print, the publishers invited me to write a similar volume up-dating the information in the earlier book.

In conformity with recent recommendations from those concerned with the education of people employed in the timber industry, less space has been devoted to wood anatomy and the microscopic details of wood structure, and greater emphasis has been given to the physical properties of wood, to composite wood products, and to the grading of timber. The timbers described are those that are commercially important today and which are likely to be available in the near future.

References to sources of further information and to relevant British Standards are given at the end of each chapter.

W.P.K. Findlay
November 1974

Acknowledgements

Permission to reproduce photographs that have appeared in publications of the Forest Products Research Laboratory has kindly been given by the Princes Risborough Laboratory of the Building Research Establishment. The letters C.C.R. after the captions of these photographs indicate that the Crown Copyright is reserved.

Figs. 8 and 34 are reproduced by permission of the British Standards Institution and the illustrations shown in Figs. 30, 31, 32 and 35 are from photographs kindly supplied by C.I.B.A.—Geigy.

The electroscan photographs of wood structure were taken in the Botany Department of the Imperial College of Science and Technology, London, and my thanks are due to Dr. John Levy for permission to reproduce them.

Finally I would like gratefully to acknowledge the great assistance given me by my wife in the preparation of the manuscript.

Contents

To
'BROOKIE'
E.H. BROOKE BOULTON

*Whose enthusiastic teaching inspired
so many to take up the study of
Timber Technology.*

CHAPTER ONE

Nature and Origin of Wood

Wood may be defined as the material that forms the trunks and branches of trees. Timber is wood cut from the trunk which can be used for constructing houses, boats, furniture and so on. The word 'lumber' is used in the United States to signify such sawn and worked timber.

As there are many thousands of different kinds of tree in the world so there is a great variety of woods, ranging from the very light, such as balsa (used to make model aeroplanes), to woods so heavy that even when dry they sink in water. Over the thousands of years that man has used wood he has learnt to choose those timbers most suitable for the purpose in hand. A tough wood was needed for his spears, for his bows elasticity was essential, while for his arrows straightness of grain was important. He learnt from experience which timbers would endure in the soil and which could be fashioned into boats that would last for decades.

In some northern countries where only a few kinds of tree grew naturally the choice of woods was very limited. In tropical forests, on the other hand, the choice was bewildering, and so, as trade between countries developed, timbers were exported from the places where the trees grew to other countries that had no forests or lacked the particular kinds of wood required. Valuable ornamental woods, such as ebony and sandalwood, were carried great distances to adorn the houses and furniture of the wealthy.

Until fairly recent times wood, and the charcoal made from it, were the principal fuels used for cooking and heating. This is still true today in the under-developed countries: in fact it has been estimated that about half the world's consumption of wood is burnt as fuel.

Traditionally woods have been classed either as 'softwoods', which means woods from coniferous trees that bear needles, or as 'hardwoods', meaning those from the broad-leaved trees. These words of course suggest that the woods from conifers are in fact softer than those from broad-leaved trees, and this was in general true of the timbers of Europe where the spruces and pines yield timbers that can be easily cut and nailed, while broad-leaved trees such as oak and beech produce harder and heavier woods. However these distinctions become blurred when one comes to consider the great range of tropical hardwoods, as the timber from many of the so-called light 'hardwoods' is in fact softer than many European 'softwoods'. Nevertheless the old terms have persisted and are still in general use in the trade. Many timber merchants in the past dealt mainly in softwoods imported from the Baltic, whilst others bought and sold hardwoods from the U.S.A., Europe and the tropical countries; but today many of the larger firms deal in both classes of wood.

Palms, which are such conspicuous trees in the tropics and in subtropical regions, do not produce true wood. Their stems do not grow in diameter as do those of other trees, and their trunks consist of tough groups of fibres in a softer material. The trunks of some palms are used locally as poles, and various ornamental articles, including walking sticks, are cut from the curiously figured tissues of their stems.

Trees grow upwards by elongation of the terminal shoots until, after perhaps hundreds of years, they reach their maximum height. This may be as much as 360 ft. At the same time their girth increases by the addition each year of a layer of new growth below the bark. The shape of the trunk depends partly on the kind of tree. Some, like birches, are naturally tall and slender, while others, such as the fantastic baobab of Africa, are fat and squat. But the form that the mature tree develops also depends very much on its surroundings. A solitary tree standing in the open will develop a short, stout stem with spreading branches, while trees grown close together in high forest produce tall, thin, straight stems as a result of their competition for light. The way trees are grown can thus affect the timber that can be obtained from them and the forester therefore attempts to grow trees that will provide the kind of timber that he expects the timber merchant will require. But the difficulty here is that in the long interval between planting and harvesting the demand may have entirely changed. Many of the oaks growing in coppices today were planted by men who thought that they would provide ready-bent limbs with which to build wooden ships, but the demand now is for long, straight stems for furniture and flooring.

When a tree is felled the butt end is found to be stouter than the upper end since many more years growth in girth have occurred at the butt. The taper of the log again depends on the way in which the tree has been grown. The less the degree of taper the more long straight planks can the sawmiller obtain from it.

When a trunk with side branches is sawn up, knots will be discovered at the points from which the branches sprang. These affect both the appearance and the strength of the wood see p. 45).

Gross Features

Annual Rings. If one examines the cross-cut end of a felled log from a tree grown in a temperate climate it will be seen that beneath the bark the wood shows, more or less sharply defined, circular rings. Each of these represents the amount of growth in girth that took place in one growing season. In Europe this is the yearly period of active growth when the trees bear leaves, hence these growth rings are called 'annual rings'.

Earlywood and latewood. Close examination under a lens of a single growth ring shows that the inner part tends to have a more open texture than the outer part where the texture is somewhat denser. This 'earlywood' (often referred to as 'spring wood' in older text books) is laid down at the beginning of the summer. As the season continues 'latewood' (formerly called 'summer wood') is formed. The widths of the growth rings vary somewhat from year to year, largely depending on the weather, and especially the rainfall, during the growing season. Generally the rings tend to narrow as the diameter of the tree increases. A simple calculation will show that it takes much more wood substance to form a ring of the same width round a large tree than round a small one. Another factor that tends to slow the rate of growth of trees as they get older is the crowding of the crown by branches of other nearby trees competing for the light.

Sapwood and Heartwood. Another conspicuous feature that appears when the cut end of the logs of certain trees (e.g. pine and oak) is examined is a belt of pale coloured wood beneath the bark surrounding the darker core. This paler outer zone is called the 'sapwood' as it is through this that the sap travels up

1. Wedge from Scots pine log; showing appearance of cross-section and of radial—and tangential—sections. Note well-marked annual rings. C.C.R.

the tree while it is growing. It contains living cells in which food materials are stored. The darker coloured central zone is known as 'heartwood' and this normally contains no living cells.

The wood from these two zones differs in certain important respects. For instance sapwood is much more permeable to liquids and is more susceptible to decay (see p. 88). But in some trees, spruce and beech for example, there is little difference in colour or properties between the inner and outer zones though the former contains no living cells.

Rays. If a log has been cut smoothly, close examination will reveal in many trees faint lines running out radially towards the bark. A few start at the very core of the tree (the pith) while others begin at various points further out. These 'rays' serve two purposes: they provide storage space for the reserve food materials (starch) that the tree stores up during the summer,

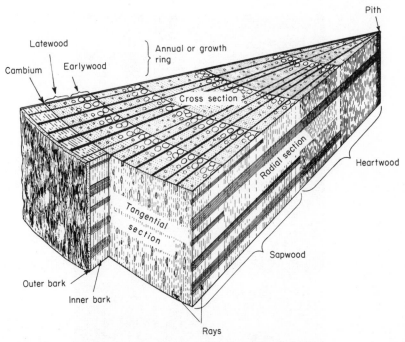

2. Diagrammatic sketch of wedge of young oak trunk; showing appearance of cross-, tangential and radial sections. C.C.R.

and they are the channels for the horizontal transport of the sap and its contents, such as the sugar that is formed in the leaves and redeposited in the ray cells after conversion into starch.

Grain. The 'grain' of the wood is, strictly speaking, the direction of the fibres in relation to the trunk. One speaks, for instance, of 'straight-grained' wood when most of the elements lie parallel to each other and to the axis of the trunk. When the fibres lie at an angle to the trunk the wood is said to be 'cross-grained'. Sometimes zones of growth lie at different angles to each other giving what is termed 'interlocked' grain. The slope of the grain has an important influence on the strength of the wood after it has been sawn (see p. 47). It also influences

3. (a) Striped figure in African mahogany caused by interlocked grain. C.C.R.
   (b) 'Silver grain' figure in radially cut oak. C.C.R.

(a)

the appearance of sawn timber giving rise to various types of 'figure'.

Figure is the term used for the pattern displayed on the surfaces of wood after it has been sawn. Interlocked grain gives a stripe or ribbon figure which is a common feature in African mahogany. Various fancy names have been given to the decorative figures that appear when woods are sawn in different ways, as for instance 'silver figure', 'blister figure', 'roe figure' and 'wavy grain' (also known as 'fiddleback mottle' because it is particularly prized when it appears in sycamore which is used for the backs of violins and cellos).

Texture. The 'texture' of wood depends on the size and distribution of the elements that compose it. Hardwoods such as oak, which have large pores, are said to have a coarse texture, while woods with very small pores, such as boxwood, have a fine texture, and are therefore suitable for engraving.

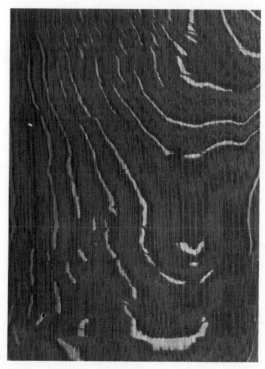

(b)

The texture of softwoods from conifers depends on the different densities of the earlywood and latewood zones. For example Douglas fir, which shows a striking difference between these two zones in each annual ring, has an uneven texture, while spruce has an even one. Only woods that have a fine, even texture are suitable for fine carving and delicate cabinet-making.

Microscopic Features

Wood is built up, like most other plant and animal tissues, of individual units called 'cells'. The young plant cell, formed by division of another living cell, is always brick-shaped. Many cells soon after their formation become modified in various ways to perform different specialised functions in the body of the plant. During the formation of wood most of the cells elongate and develop thickened walls.

The actively-growing tissue in a tree is found only in the points at the tips of the main trunk and branches, and in the thin layer between the wood of the trunk and the bark. This layer, which is called the 'cambium', is only a few cells thick. These cells divide by forming vertical walls which make, on the inside, new elements of the 'xylem' (or wood) and, on the outer side, new elements of the 'phloem' (or bast). The main functions of the xylem are to conduct water up from the roots and to provide the strength to support the tree. Its elements are therefore elongated and have stiff, thickened walls. The function of the phloem, on the other hand, is to conduct the substances synthesised in the leaves down to the roots and trunks. It is always quite a narrow zone of rather thin-walled active cells and is removed with the bark when the tree is sawn up into timber. Outside the phloem is the bark with its own active growing layer or 'phellogen' that provides for the increase in the thickness of the bark as the tree increases in girth. The cells of the bark are empty of contents and their walls are impregnated with a waxy substance which makes them resistant to the passage of moisture. This property also enables the bark to act as an effective insulating layer against extremes of heat and cold and protects the underlying wood against drying out. Natural cork is prepared from the thick bark of the cork oak and its use for making stoppers for bottles has been known for centuries.

The fine structure of wood can be studied in three ways:

(i) By breaking the wood up into its individual cells by a process known as 'maceration'. This is effected by boiling thin shavings of the wood in a solution of potassium chlorate in 50 per cent nitric acid and then teasing the fibres apart with needles. After being stained and mounted, they can then be observed under a microscope.

(ii) By cutting very thin sections of the wood which are stained and then observed microscopically by transmitted light. Sections cut across the grain are called transverse while the sections cut along it are called longitudinal. The latter can be cut in two ways, either radially or tangentially to the axis of the trunk. The appearances of these three types of section are completely different and it is necessary to study sections of each type in order to understand fully the structure of any wood.

(iii) By examining small, specially prepared blocks of wood by means of an electroscanning microscope. This gives a three-dimensional picture of the fine structure under very high magnification (see Figs. 4 and 5).

The microscopic features of softwoods from conifers and of hardwoods from broad-leaved trees are so different that they must be considered separately.

Structure of Softwoods

The microscopic structure of the softwoods from conifers (*Gymnosperms*) is much simpler than that of the wood of the broad-leaved trees (*Angiosperms*). When the wood of the former is macerated i.e. broken-up into its individual units, these will almost all appear as long, thin, needle- or cigar-shaped cells with round markings on their wall. These are known as 'tracheids' (see Fig. 6). They are usually about 2·5 to 5 mm long, occasionally reaching 10 mm. Tracheids with thin walls come from the earlywood zones and function as conducting tubes for the sap. The thicker-walled tracheids are from the latewood and contribute more to the strength of the wood. The

4. Electroscan photograph of small block of Scots pine, highly magnified. Top surface: cross-section showing resin ducts (the two large pores), the earlywood (dark) and latewood (light) zones, and rays (appearing as bold, spaced lines against the cellular background). Right surface: radial section, showing two rays (horizontal rows of cells). Left surface: tangential section.

round markings on the walls are the valve-like openings between one tracheid and another. Their rather complicated structure can be more easily studied in sections of the wood. There will also be seen a small number of thin-walled rectangular cells derived from the rays and resin ducts. These cells are known as 'parenchyma' cells.

In Fig. 7 the transverse section shows the annual growth rings with the tracheids cut across so that the thickness of their walls can be observed—the earlywood tracheids having thin walls and the latewood ones rather thicker. Most of the

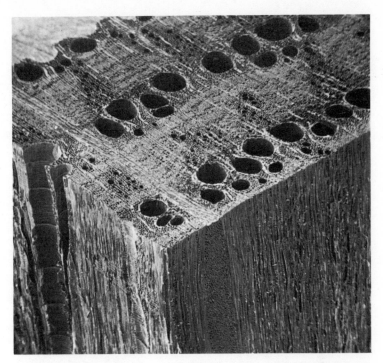

5. Electroscan photograph of two annual rings of oak, highly magnified. Top surface: cross-section showing large vessels in earlywood separated by fibres and small vessels in latewood. A large ray can be seen parallel to the left edge of the top surface and in section on the right (tangential) surface. Left surface: radial section.

rays are only one cell wide but a few are wider at the centre and taper off at top and bottom (see Fig. 7).

In many conifers, e.g. pines, spruces and larches, there are long resin ducts which are lined with thin-walled parenchyma cells that secrete resin. These form a useful diagnostic feature enabling different groups of conifers to be distinguished.

The radial longitudinal section (Fig. 4) shows the tracheids dovetailing into each other. The thin-walled tracheids of the earlywood show many bordered pits, while the thick-walled latewood tracheids have only a few small simple pits. The bordered pit is in essence a valve-like opening between adjacent tracheids. It serves a useful purpose in preventing flow of water

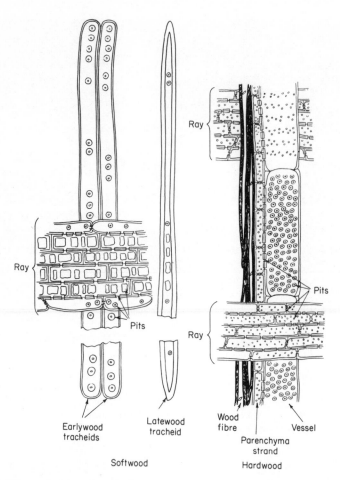

Ray

Ray

Ray

Pits

Pits

Earlywood
tracheids

Latewood
tracheid

Wood
fibre

Vessel

Parenchyma
strand

Softwood

Hardwood

6. The cells of which wood is composed. C.C.R.

into empty non-functioning adjacent cells. The pit membrane
is thickened at its centre into the round 'torus' which is
suspended by a network of fibrils that permit liquid to pass
through the spaces between them. When the pressure changes
on one side of the bordered pit the torus is drawn across
sealing the opening. Simple pits are openings in the wall without this valve-like mechanism. The rays in this radial longitudinal section appear as bands of tissue some 4 to 10 cells in
height and consist of thin-walled elongated brick-shaped cells.

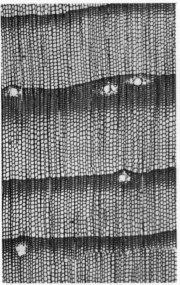

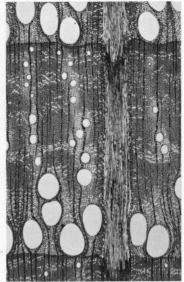

7. a) Cross-section of pine showing four growth rings with resin ducts in the latewood. × 30 C.C.R.

b) Cross-section of a ring-porous wood (oak) with wide and narrow rays. × 30 C.C.R.

c) Cross-section of diffuse porous wood (beech) with wide and narrow rays. × 30 C.C.R.

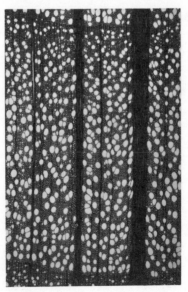

The tangential longitudinal section (Fig. 1) may of course be cut through a band of earlywood or of latewood, so that the appearances of different consecutive sections cut tangentially through the same piece vary appreciably. The cut ends of rays form a conspicuous feature in the tangential section. In a few of the larger of these there may be a resin duct running horizontally in the tree.

Structure of Hardwoods

The structure of hardwoods is more complex than that of softwoods and the elements that compose the wood are more varied.

In hardwoods the conduction of sap takes place through long tubes known as 'vessels', while the mechanical strength is provided by fibres. Vessels are formed from a series of elongated cells, one above another, in which the end walls largely disappear, giving a similar effect to a drain pipe made up of individual sections (see Fig. 6). In cross-section these appear as round openings or 'pores'. These pores vary greatly in size. In some woods, such as oak, that have a coarse texture they are just visible to the naked eye, in others they are microscopic. In species such as beech vessels of more or less the same diameter are formed throughout the growing season (see Fig. 7). These are called 'diffuse porous' woods. Many tropical timbers from trees that grow throughout the year are of this type. In other species, such as oak, the vessels formed in the spring are much wider than those formed later in the growing season and distinct rings of pores are visible to the naked eye. Such woods are called 'ring porous' and in these trees it is quite easy to count the number of annual rings, and hence to know the age, by examining a cross-section of the log or stump.

Most of the mechanical strength of the wood is derived from the fibres, which are long, narrow, thick-walled cells with pointed ends. Therefore the density and strength of a timber depends to a large extent on the quantity of fibres it contains. Some hardwoods, oak for example, contain tracheids similar to those in softwoods and while these act as conducting tubes they also contribute to the strength of the wood. In addition there is in the wood of broad-leaved trees more storage tissue than in softwoods. This takes the form of thin-walled 'parenchyma' much of which is in the rays which are generally wider than those of softwoods.

The wood parenchyma may be arranged in a variety of ways either scattered throughout the fibrous tissue or associated with the vessels. Sometimes it forms broad conspicuous bands which are visible to the naked eye as lighter-coloured zones.

The rays in many hardwoods, which are much more conspicuous than in softwoods, may occupy a fifth or more of the total volume of the wood. While the rays in softwoods are usually only one cell wide, in the majority of hardwoods they are 'multivariate', that is to say, more than one cell in width. Timbers like oak and beech that have very wide rays tend to split easily along the grain, but the rays give a pleasing figure to wood that has been cleft or sawn radially. This type of figure is often referred to as 'silver grain'.

Structure of Sapwood and Heartwood

Only minor changes occur in the anatomical structure of the wood when sapwood changes into heartwood. A feature of this change is the entry of air into the tracheids and vessels, replacing the sap that had previously flowed through the active sapwood. As a result of this the cells in the wood parenchyma and in the rays die and no longer store reserve food materials. Microscopic tests fail therefore to reveal any starch in the cells of heartwood. In some woods, notably the white oaks (English oak) the thin-walled, bladder-like extensions from the parenchyma cells grow out into the vessels completely blocking them. These 'tyloses' thus render the wood very resistant to the passage of liquids, making the woods that contain them very suitable for beer barrels and wine casks.

In some woods, notably some species of eucalyptus, the cells of the heartwood become filled with a dark, reddish-brown gum, and in others, such as makoré and turpentine, the ray cells may contain so much silica that saws are rapidly blunted when cutting planks.

Great progress has been made in the study of wood anatomy

in the past half century and important contributions to the study of wood structure have come from Timber Research Laboratories in England, France, the U.S.A., Australia, India, Malaya and elsewhere. Wood anatomists have formed an international association one of whose achievements has been the publication of a multilingual glossary of terms used in wood anatomy.

The bibliography to this Chapter lists some of the many publications which the student can consult if he wants to study this subject more deeply.

BIBLIOGRAPHY

Desch, H. E. *Timber: Its Structure and Properties* 4th edition. (London, Macmillan), 1973.

Eames, A. J. and MacDaniels, L. H. *An Introduction to Plant Anatomy* 2nd edition. (New York and London, McGraw Hill) 1947.

Jane, F. W. *The Structure of Wood* 2nd edition. (London, A. & C. Black) 1970.

Zimmermann, M. H. (editor). *The Formation of Wood in Forest Trees.* (New York and London, Academic Press), 1964.

Rendle, B. J. (revised, Brazier, J. D.) *The Growth and Structure of Wood.* Forest Products Research Bulletin No. 56 (London, H.M.S.O.), 1971.

U.S. Forest Products Laboratory. *Wood Handbook.* (Washington), 1955.

CHAPTER TWO

Conversion of Logs into Timber

When trees are felled it is sometimes known precisely what use will be made of them, as when small, young trees are removed during the thinning of plantations. But with larger trees the sawmiller must use his judgement in deciding how best to convert a log to obtain the maximum yield of reliable timber cut into the sizes that he anticipates his customers will require.

Trees with a butt diameter of less than 9 in (23 cm) are seldom sawn up in this country but are cross-cut and used either as posts or as pit-props for supporting the roofs of mines. Large quantities of the smaller thinnings go straight to the pulp mills to be made into paper or chipboard. It is estimated that by 1975 rather less than half the consumption of wood in North America and Europe will be sawn, as the greater quantity will be used to make paper and panel products.

Trees rather larger than the first thinnings are removed in subsequent thinnings and these provide poles used for carrying overhead telephone lines and electricity supply cables. Today these are usually of only moderate size as the main telephone lines that formerly required very large poles are now carried underground. Posts and poles that have to be impregnated with a wood preservative before they are used must have the bark removed and be stacked to season. Ingenious machines which scrape off the bark, even though the logs are uneven or twisted, have been devised, so the tedious task of removing the bark with a draw knife is seldom undertaken nowadays in the developed countries. Formerly all this bark was either burnt or just left to rot, but recent studies have shown that if it is pulverised, it can be used in place of peat, and it is useful as a mulch to conserve moisture and smother weeds amongst shrubs.

Conversion of felled trees and logs into usable timbers is carried out in a number of different ways, most of which are very ancient and some now almost obsolete. The main processes are hewing, cleaving, slicing, peeling and sawing.

Hewing. Before the introduction of mechanically driven saws the easiest way to produce a square beam from a round log was to chop away the outer layers on four sides. For this work a broad axe with a blade tapered on one side only was used. For finer work the hewer used an adze, which is essentially an axe with a curved blade set inwards at right angles to the handle like a draw hoe. A worker skilled in the use of the adze could produce an almost perfectly smooth surface, as can be seen in the beams of many old houses and the tops of old refectory tables. Hollow wooden vessels such as pig troughs were sometimes hewn out of the solid and the adze was also used to hollow out the elm seats of Windsor chairs made in the Chilterns. Nowadays the adze is used in the shaping of large timbers for boat-building and in the tropics for squaring logs of mahogany and similar timbers.

Cleaving. Some kinds of straight-grained wood can readily be split, especially in a radial direction, and the art of cleaving wood to produce thin material from logs is an ancient one. Cleft material is often stronger than sawn pieces of the same size because the cleft pole or slat follows the grain of the wood which is nowhere cut across.

Cleaving is a process that is not easily mechanised and therefore many of the woodland crafts (so delightfully described in H. L. Edlin's book) have largely died out, the cost of labour having increased so much as to make the work no longer profitable. Cleft chestnut paling is however still in demand, as the wood of the sweet chestnut, even from small logs, is naturally very durable. Cleavers prefer poles that are

4–5 in thick at 6 ft from the ground, a size that they reach on the coppiced trees after 10 or 12 years. Some 25 000 pales, enough for a mile of fencing, can be obtained from an acre of chestnut coppice.

Trackway paling, used to provide a rough road for lorries over sandy ground, was made in great quantities for the invasion of the Normandy beaches in 1944.

The best ash handles are made from cleft pieces of the wood in which the grain is undamaged. The final shape is achieved with the use of a spokeshave or a draw knife.

Slicing is the term applied to the process of cutting thin slices of wood with a knife moving to and fro across the long axis. It is the method generally used for preparing decorative veneers that display the grain of the wood to the best advantage.

If the wood to be cut is a hard one, the baulk of timber may be steamed or heated in hot water to soften it before cutting. If the wood has been heated throughout to about 88°C (190°F) boards up to half an inch thick can be cut by slicing. An added advantage is that no timber is wasted as sawdust.

Peeling. The preparation of thin veneers by turning the log against a fixed knife is known as peeling. It is the process used in the production of plywood (see p. 173).

Sawing. The origin of the sharp toothed metal saw is lost in history but it is known to have been used in Britain before the coming of the Romans. At first it was used mainly for cutting timber across the grain, the squaring up of logs being done mainly with the broadaxe and adze. For sawing logs by hand into boards the log was supported above a saw pit and moved forward as the sawing proceeded. Two men were needed to work the saw. The top 'sawyer' stood on the log and lifted the saw for each stroke and the 'pitman' pulled it down. A sawpit of this kind was still in use in Essex as late as 1948, and pit sawing is still practised in parts of Africa.

Sawmills driven by water power were in use in mediaeval times, and I have myself seen one in Scotland since the last war. The circular saw was patented in 1777 by one Samuel Miller and this idea was taken up by Brunel who, in 1799, started the first steam-driven sawmill at Chatham Dockyards. By 1824 circular saws were in regular use but since then they have been improved and developed in many ways. They have now been superseded in many mills by the bandsaw. This has a steel band with teeth along one edge revolving around two wheels. The simple circular saw, however, is still widely used in portable sawmills in the woods, and it is the main tool for the cross-cutting of wood. In Northern Europe logs are generally converted by 'frame saws' consisting of a vertically reciprocating frame in which a number of saw blades are mounted.

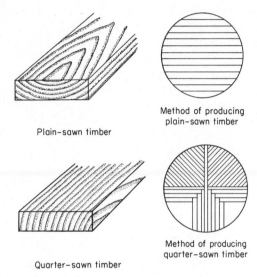

Plain-sawn timber

Method of producing
plain-sawn timber

Quarter-sawn timber

Method of producing
quarter-sawn timber

8. Methods of sawing logs to produce plain and quarter sawn timber.

Logs can be sawn up into boards or planks in either the tangential or the radial direction. The simplest way is to saw the log straight through from one side to the other, thus obtaining a number of tangentially cut boards and one radially cut including the centre of the tree. Boards cut in this way are said to be 'plain' or 'flat' sawn.

The alternative is to saw up the log so that every board is cut more or less in a radial direction; that is to say, the rings meet the surface of the board at an angle of more than 45°. Boards cut in this way are said to be 'quarter' sawn.

In some trees the central core of the trunk is much weaker and less durable than the rest of the heartwood and it may therefore be desirable to 'box the heart' and use the core wood only for rough work, pulp wood or fuel.

Plain sawn timber is cheaper to produce than quarter sawn but the boards tend to cup more, and if used for flooring the softer parts of the annual rings tend to wear away so that the rings may 'shell out' and this may lead to splinters forming on the surface. Quarter sawn wood shrinks and wears more evenly and if the timber is one with wide rays such as oak it displays a much more attractive figure.

When a log is sawn through and through the planks are sometimes reassembled into the original form of the log for seasoning. This is then known as a 'boule'. The term is generally applied to European timber such as oak. The original rounded surface of the tree remaining in a piece of such converted timber is known as 'wane'.

After conversion and seasoning the timber may either be used as it is, or re-sawn into smaller sizes and sold as 'rough sawn', or as 'prepared stock' after it has been planed. Such planed timber is often said to be 'wrot', or in Scotland 'dressed'.

The technology of sawmilling and the maintenance of saws are subjects on which many volumes have been written and for further information reference should be made to those listed in the bibliography.

Sizes and Definitions of Sawn Timber

The timber trade has recently changed from Imperial units to metric units and in the United Kingdom timber sizes and quantities are now quoted in metric terms. The following are the names of the sizes most commonly used and defined in BS 565 of 1972.*

Batten   A piece of square sawn softwood, 50–100 mm thick and 100–200 mm wide.

Baulk   A piece of square sawn or hewn softwood, approximately square in cross-section and of greater size than 100 mm × 125 mm.

---

*Glossary of terms relating to timber and woodwork, British Standards Institution, 2 Park St., London W.1. (from whom copies of this standard may be obtained).

Board   a) Softwood. A piece of square sawn timber under 50 mm thick, and 100 mm or more wide.
b) Hardwood. A piece of square sawn or unedged timber, 50 mm or less thick, and 150 mm or more wide.

Deal   A piece of square sawn timber softwood 50–100 mm thick and 225–300 mm wide.

Lath   A piece of sawn or cleft timber 6–17 mm thick and 22–30 mm wide.

Plank   a) Softwood. A piece of square sawn timber 50–100 mm thick and 250 mm or more wide.
b) Hardwood. A piece of square sawn or unedged timber over 50 mm thick and of various widths.

Scantling   a) Softwood. A piece of square sawn timber 50–100 mm thick and 50–125 mm wide.
b) Hardwood. Timber cut to an agreed specification such as waggon oak scantling; or a square edged piece not conforming to other standard terms.

Stave   A piece of sawn or cleft timber intended to form part of a cask.

Strip   a) Softwood. A piece of square sawn timber less than 50 mm thick and under 100 mm wide.
b) Hardwood. A piece of square sawn timber usually 50 mm or less thick and 50–140 mm wide. Usually for flooring.

Measures. Formerly softwood was generally sold by the Petrograd Standard which is 165 cu ft ($= 4.67$ m³) and hardwood by the cubic foot. Since metrication timber is bought and sold by the cubic metre. A board metre is the amount of timber equivalent to a piece 1 m² and 25 mm thick ($= 1/40$ m³).

BIBLIOGRAPHY

British Standards Institution. *Glossary of Terms relating to Timber and Woodwork.* BS 565, 1972.

Edlin, H. L. *Woodland Crafts in Britain. An Account of the Traditional Uses of Trees and Timber in the British Countryside.* (Newton Abbot, David & Charles), 1973.

Goodman, W. L. *The History of Woodworking Tools.* (London, G. Bell), 1964.

Hayward, C. H. *Tools for Woodwork* revised edition. (London, Evans Bros.), 1973.
Hudson, H. R. *Woodworking Machinery* 3rd edition. (London, Evans Bros.), 1960.
Serry, V. *British Sawmilling Practice*. (London), 1963.

CHAPTER THREE

Physical Properties of Wood: General

The uses to which a particular timber can be put depend in the first place on its physical properties. We think of using wood when we require a material that is strong in proportion to its weight and which can be cut, nailed, and screwed with ease. We may also want a material that is pleasing to the eye and warm to the touch. All these qualities, with the exception of the purely aesthetic one, are due to physical properties that can be measured.

Woods vary greatly in these physical properties. The wood from one species of tree varies considerably according not only to the rate of growth of the tree from which it was cut, but even from what part of the trunk it was taken. It must always be remembered, therefore, that any figures quoted for the physical properties of a timber are only the average values. It is useful to know the range of variation that one may expect to find as it may be necessary to allow for the inclusion of pieces of wood much below, or much above, the average when calculating the dimensions of members to be used in a structure.

Density of Wood

The density of a material is the mass of a unit volume of that material. The density of wood used always to be expressed in

Imperial measures as the weight in pounds of a cubic foot of timber. In metric terms it is expressed as kilogrammes per cubic metre.

The density depends to a considerable degree on the amount of moisture held in the wood. Freshly felled timber full of sap is much heavier than the same timber after it has been seasoned (dried). With experience one can roughly estimate the moisture content of a piece of familiar wood by assessing its weight in the hand. It is therefore usual when describing a timber to quote its weight when green (freshly sawn out of the log) as well as its weight when seasoned down to 12 per cent moisture content.

Different timbers vary enormously in their density, from about 5 to 70 lbs per cu ft. However, the basic wood substance that composes the cell walls always has about the same specific gravity in all woods, that is, about 1·5, so that the variations in density reflect the amount of cell-wall substance relative to the amount of air space in the wood.

Woods may roughly be grouped into five classes on the basis of their densities thus:

Class	kg/m³	lb/cu ft	Examples
Very light	80–240	5–15	Balsa
Light	400–560	25–35	Spruce
Medium	560–720	35–45	Ash
Heavy	720–960	45–60	Jarrah
Very heavy	Over 960	Over 60	Lignum Vitae

Specific gravity is the ratio of the weight of a substance to that of an equal volume of water. As the volume of a piece of wood alters with changes in moisture content it is usual to adopt a nominal specific gravity for wood which is based on the weight when oven dried and the volume at the time of test.

In some trees, particularly in ring-porous hardwoods, the density may vary greatly from tree to tree according to the rate of growth which determines the relative amounts of the porous earlywood and of the latewood which consists mainly of fibres. Ash, for instance, may vary from 33–52 lb/cu ft (528–832 kg/m³).

Density has an important influence on the strength of wood. Broadly speaking the denser a piece of wood the stronger it is. However, the strength may be assessed in a variety of ways see p. 33) and some strength properties are more closely related to density than others. Hardness is probably the quality most closely linked to density.

There are certain purposes for which heavy woods are definitely unsuitable, for example for the decking of ships where strength and durability are important but where, in the upper decks, additional weight is undesirable. Very heavy woods are generally used for purposes for which their other associated qualities, such as hardness, make them particularly suitable; for some specialised uses, such as balls for croquet and bowls, their weight is an actual advantage.

Very dense woods tend to be impermeable as there is little space into which liquids can penetrate. It is therefore not possible to treat them effectively with wood preservatives. However as most of them are very durable, treatment is seldom necessary.

Very light woods, such as balsa, find a use in the construction of model aircraft. The early air planes were built of Sitka spruce, a wood that is very tough for its low density. Packing cases too require light-weight woods which at the same time are strong enough to protect their contents.

Colloidal Nature of Wood

Wood is a colloidal substance which always has a certain affinity for moisture. Only when it has been dried in an oven at a temperature slightly above the boiling point of water is it ever completely dry. Normally it continually absorbs or loses moisture from the atmosphere as the humidity of the air around it changes. As the moisture content rises or falls, the wood swells or shrinks. For this reason the precise control of moisture content is one of the most important aspects of processing timber for any purpose in which exact dimensions of the finished article are important (see Chapter 6).

Colour of Wood

The sapwood of most timbers is pale in colour, generally cream or light buff though almost white in some woods, but the colour often darkens somewhat after exposure to the air.

The heartwood when it develops is distinguished by its much darker colour. In many conifers, such as Douglas fir for instance, it has a reddish-brown colour. This dark colouring is generally due to the production of tannin and similar substances by the cells of the sapwood as they die. These substances confer some degree of resistance to decay on the wood (see p. 88).

Sometimes colour develops irregularly giving rise to stripes and bands of a darker shade which may considerably enhance the beauty of the wood. Walnut is one such that is much sought after for furniture. A few woods, such as ebony, have an intensely dark, almost black, heartwood. Others, notably purpleheart, are bright purple when freshly sawn, but the colour darkens to a deep crimson brown on ageing.

The differences in colour between various woods tend to become less pronounced if the sawn surface is exposed to the light and air, dark-coloured woods tending to fade and pale ones to darken. Wood should therefore be coated with a transparent varnish or seal if it is desired to retain the original appearance (see p. 103).

Very light-coloured woods are valued for making objects, such as pastry boards or rolling pins, that are to come into contact with food. Fashions for light-coloured furniture come and go. At one time oak treated with lime to lighten its colour was very popular for dining room and other furniture. In an earlier generation oak fumed with ammonia, which darkened the natural colour of the wood, was very fashionable.

Abnormal colourations are often an indication of incipient decay (see p. 83). Sometimes they take the form of narrow, dark zone-lines which look as if they had been drawn with a black pencil. A bluish-grey discolouration in the sapwood of pines is usually caused by the growth of moulds through the wood (see p. 192) but a blue-black stain in oak is due to a chemical reaction between tannins in the wood and some source of iron. Discolouration of hardwoods may also be the result of infection by the moulds that cause the so-called 'bluestain' (see p. 192).

Perhaps the best-known example of a fungal discolouration that actually enhances the value of the wood is 'Brown Oak'. This is caused by the growth of the Beefsteak Fungus through the trunk of oak trees. 'Green Oak' is also the result of a fungus which grows through fallen branches of the tree staining them a vivid green colour. These variously-coloured woods are much prized for marquetry work (see Chap. 17).

The lustre which gives a bright appearance to such woods as satinwood is the result of the cell walls reflecting the light, and the alternate light and dark appearance of wood blocks laid in herring-bone fashion illustrates how the appearance of wood may depend on reflected light. Wood that has been cut radially (i.e. quarter sawn) is generally more lustrous than flat sawn wood.

Heat Conductivity

The ability of any material to conduct heat depends on its specific conductivity and its specific heat; i.e. the amount of heat required to raise unit weight (1 gm) of its substance 1°C. Though the specific heat of wood substance is similar to that of many metals, or even greater, the conductivity is very much less because all dry woods contain a considerable volume of air enclosed in the cells. The conductivity of green unseasoned timber is much higher than that of dry wood as in green wood the cells are filled with water which is a better conductor of heat than is air.

The conductivity of dry wood depends greatly on its density; the lighter the wood the better are its insulating properties. Balsa wood was used as an insulating material long before the invention of expanded plastics.

The excellent insulating properties of timber make it a particularly suitable material for walls of buildings in cold climates. The traditional cottages in Russia were built of solid logs. Panelling a room with wood not only greatly reduces the heat loss through the walls but it also enables the room to be heated more quickly. Because wood conducts heat so poorly it is the ideal material from which to make such things as saucepan handles, and for the same reason silver teapots are often made with an insert of ebony between the pot and the handle so that the latter does not become uncomfortably hot to hold.

The slow conduction of heat through timbers renders it

suitable (rather surprisingly at first thought) for the construction of fire-resistant doors. Provided that the panels do not separate and so make openings through which flames can pass, a well-made, solid timber door can resist the penetration of fire for a long time. For the same reason large timber beams may continue to support a building long after unprotected steel joists have collapsed due to softening of the metal by heat. Similarly, papers in a metal filing cabinet are more likely to be charred than those in a wooden one during a fire of fairly short duration.

All substances tend to expand when heated and, provided the moisture content is kept constant, this is true also of wood; but under practical conditions heating wood always tends to dry it out and so, as wood shrinks on drying, the net result is usually for wood to shrink when it is heated.

Electrical Conductivity

Dry wood is a poor conductor of electricity but it becomes a partial conductor when it contains moisture. The degree of conductivity is closely related to the amount of moisture in the wood and this property has been used in many electric moisture meters for wood to assess the moisture content (see p. 66).

Calorific Value of Wood

Wood, and charcoal prepared from wood, are still the principal fuels in many parts of the world. When forests were thought to be almost inexhaustible and timber was cheap, wood was even used to fuel the railway locomotives in many parts of Africa.

The calorific value of absolutely dry wood is only about 60 per cent of that of an equivalent weight of coal, but as firewood in practice always contains some moisture its value as fuel is in fact even less than this. Wood for burning should, of course, always be as dry as possible as much of the heat produced by combustion may be wasted in boiling off the water from wet wood. Elm is almost unburnable in a green condition, and though some woods, such as holly and ash, that have a lower moisture content when growing can be burnt unseasoned it is wasteful to do so and they should always be dried first. Knotty softwoods often give off sparks explosively and if used on an open fire a guard is necessary. The best fuel woods are hardwoods such as beech, oak and sycamore.

Wood fuels produce far less ash than does coal and the ash they do make is rich in potash and phosphates and therefore useful as a fertiliser, especially on poor forest soils that lack these elements.

Charcoal is made by partial combustion of wood under a very restricted supply of air so that only carbon remains, and the volatile compounds that give rise to flames when burnt are all drawn off. Since charcoal burns without flame or smoke it is very suitable for cooking and is still used for barbecue grilling. Formerly charcoal was used in the manufacture of iron and the destruction of the great oak forests in the Weald of Kent and Sussex during the 18th century was largely due to the huge demand for charcoal for the Sussex iron works.

Wood for fuel should always be purchased by volume, unless a strict specification for its moisture content is laid down. Otherwise there is a temptation for the vendor to sell it in as wet, and therefore as heavy, a condition as possible. Fuelwood has traditionally been sold by the cord (a volume equal to 8 × 4 × 4 feet,) in this country and in France by the stere which equals a cubic metre.

During the Second World War when petrol was scarce in many countries wood was used as a source of fuel for motor engines. By destructive distillation the greater part of the potential energy of wood can be converted into so-called 'producer gas', a flammable mixture consisting mainly of carbon monoxide and hydrogen. In Germany in the late thirties it was not uncommon to see filling stations supplying wood fuel for lorries (*Holztankstelle*).

Acoustic Properties

Wood is used for many musical instruments as it is a very resonant material that vibrates when activated by sound waves to give a pleasing sound. One of the simplest uses is in the xylophone where strips of a dense wood of differing lengths are struck to give notes of differing pitch.

Uniformity of texture resulting from an even growth, and freedom from knots, are the important qualities sought for in timber to be used for musical instruments. The sounding board of a piano is generally made of finest quality spruce, and this wood, usually from Rumania, has also traditionally been the one most prized for making the belly of violins, while maple is used for the back (see p. 215). For wind instruments stability under varying conditions is a most important quality. Cocuswood from the West Indies was the timber of choice for clarinets etc. on account of its fine uniform texture, oily nature, and handsome dark appearance.

If wood is fixed firmly so that it cannot vibrate it absorbs sound waves and does not reflect them. For this reason its use as panelling in concert halls reduces echoes and improves acoustics. It was also largely for this reason that wood blocks were laid in the streets of London and other large towns in the days when the noise came mostly from the iron-shod hooves of horses before the coming of the internal combustion engine.

Odour and Taste

Many woods have a characteristic smell which is usually most pronounced when the timber is freshly sawn. These odours are often a great help in identification but it may be necessary to expose a fresh surface in order to catch them. Most softwoods, particularly the pines, have a resinous smell, and Western red cedar can always be recognised by its aromatic odour. A few woods, such as Queensland walnut, have a really foul smell when freshly cut. It is not unusual for logs that have lain for some time in a log pond to have a very high smell, but this is due to bacterial growth in the water. These bacteria enter the wood and by their action render it more permeable to preservatives (see p. 121).

Unusual odours may result from fungal infections. Perhaps the most extraordinary case of this kind was a disease in some Persian oak which was used to make beer barrels after the war. The wood was infected with a wood-rotting fungus that smelt strongly of cheap perfume and made beer stored in any infected casks quite undrinkable.

Some woods have found special uses on account of their particular odour. For instance cigar-box-cedar has a pleasant aroma which was considered to enhance that of cigars packed therein.

The essential oils that give rise to odours sometimes make timbers repellent to insects such as clothes-moths and so chests for storing clothes used to be made of camphor wood or cedar wood. Though such timbers are often repellent to termites they may lose their resistance as the oils evaporate.

Absence of smell is important for any wood that is to be used for packing food-stuffs or tobacco that may pick up volatile substances and thereby become quite unsaleable.

The taste of woods is generally undesirable so wooden objects, such as chopping boards, that are to come into contact with food must be made of a kind that will not impart any flavour. However certain substances that come from tree trunks are eaten and enjoyed, such as maple syrup from the Canadian maple tree, and chewing gum (chicle) from the sapodilla plum. There is also a timber called waika chewstick which certainly suggests that it may have a pleasant flavour, at least to some palates.

CHAPTER FOUR

Physical Properties of Wood: Mechanical

Timber is used for many purposes in which its strength is a matter of critical importance. As a result of trial and error over many centuries practical builders of houses and boats had arrived at certain sizes of familiar timbers that were adequate for the purposes in hand. However with the advent of new timbers of unknown strength it became necessary to carry out precise tests to find out just how strong they were. At the same time the price of timber continued to rise and the need arose to use it more economically and for architects and engineers to be able to calculate exactly the optimum dimensions necessary for different structural members.

Unlike homogeneous materials (such as metals and plastics) wood is an anisotropic material, which means that its strength properties differ greatly according to the direction of the grain in relation to the stress applied. So when referring to the strength properties of a timber it is always necessary to specify whether the property refers to a stress applied parallel to the grain or at right angles to it. The strength properties of sheets of wood can be rendered uniform in two dimensions by gluing together thin sheets of veneer with the grains running at right angles. This gives the material known as plywood.

The significance of different types of strength tests depends on the purpose for which the timber is to be used. Thus: *Bending Strength* is important in floor joists which must resist stresses imposed by the loads which the floors have to carry. *Stiffness* (modulus of elasticity) is the ability of a material to resist bending. It is an important property in joists, otherwise the ceilings below them would crack if the floor above flexed under load. It is also a critical property in determining the relative strengths of a long column or strut and a short one.

Tensile strength is necessary for timbers that are to be bent into curved shapes after steaming.

Toughness, the ability to resist suddenly applied stresses, is required for tool handles, shunting poles, and many sports goods such as hockey sticks.

Compressive strength is required for pit props and stanchions that have to bear loads imposed on them parallel to the grain of the wood.

Hardness, or resistance to indentation, is desirable in flooring, decking, mallets, rollers, bearing blocks etc.

Cleavability or resistance to splitting may be a desirable property in packing case timbers that have to be nailed, but a disadvantage if it is wished to cleave the wood as for instance in making cleft shingles for roofing or pales for fencing. Also when preparing firewood it is convenient to be able easily to split large branch wood and logs.

TABLE I.

Species of wood	Condition	Density	Nominal sp.gr.	Static bending			
				Max. bending strength	Modulus of elasticity	Energy consumed	
						to max.load	to fracture
		kg/m^3		N/mm^2	N/mm^2	mmN/mm^2	mmN/mm^2
Baltic redwood	Green	625	0·41	44	7 700	0·070	0·186
	Air-dried	481	0·43	83	10 000	0·098	0·120
Douglas fir	Green	673	0·45	54	10 400	0·059	0·164
	Air-dried	545	0·49	93	12 700	0·088	0·192
Ash	Green	801	0·53	66	9 500	0·172	0·375
	Air-dried	689	0·60	116	11 900	0·182	0·281
European oak	Green	833	0·56	59	8 300	0·079	0·178
	Air-dried	689	0·61	97	10 100	0·093	0·167
Balsa	Air-dried	176	0·16	23	3 200	0·018	0·035

The ways in which woods are tested for these properties are given below and some of the results are shown in Table 1.

Testing the Strength of Timber

Formerly strength values of timbers were expressed in Imperial Units, but since the timber trade has now adopted the metric system all data are expressed in SI metric units, as follows:

Force $1 N = 0.225$ lbf
Force/length $1 N/mm = 5.71$ lbf/in
Stress $1 N/mm^2 = 145.0$ lbf/in^2
Work per unit volume 1 mmN/mm$^3 = 145$ in lbf/in^3

The SI unit, known as the newton, is defined as the force that induces in 1 kg an acceleration of 1 metre/second/second.

Bending

The strength properties that can be determined from a static bending test in which a load is imposed at a constant slow rate

Impact	Compression	Hardness	Shear	Cleavage
Shock resistance Max. drop	Parallel to grain N/mm^2	On side grain N	Parallel to grain N/mm^2	Average of radial and tangential N/mm width
0.69	21.0	2 490	6.3	10.5
0.66	45.0	3 690	137	10.5
0.66	25.9	2 140	7.2	8.8
0.86	52.1	2 980	10.8	8.9
1.17	27.2	4 270	9.0	
1.07	53.3	6 140	16.6	
0.86	27.6	4 670	9.1	17.4
0.84	51.6	5 470	13.7	17.3
	15.5		2.4	

(6·6 cm per minute) at the centre of a small beam, 30 × 2 × 2 cm are:

Modulus of rupture, which is the equivalent stress in the outermost fibres at the point of failure.

Modulus of elasticity, which expresses the relation between stress and strain and is of importance in determining the deflection of a beam under load.

Work to maximum load, which is a measure of the toughness of the timber under bending stress.

Impact Bending

In this test a specimen of the same size as that used in the static bending test is exposed to repeated blows from a hammer weighing 1·5 kg which is raised stepwise until the beam breaks or is deflected 60 mm. The maximum drop of the hammer is a good measure of the toughness of the timber or its shock resistance to a suddenly applied load.

Compression parallel to the grain

In this test the load is applied to the end of a specimen 6 × 2 × 2 cm in size at the rate of 0·6 mm a minute.

Hardness

Hardness, or resistance to indentation, is measured by driving a specially hardened steel tool rounded to a diameter of 11·3 mm (having a projected area of 100 mm²) to a depth equal to half its diameter. The test is carried out on both the radial and tangential faces of the 2 × 2 cm test piece.

Cleavability

This is measured by pulling apart the two halves of a piece 4·5 × 2 × 2 cm (see Fig. 14).

Shear strength

Is measured by applying a load to half the width of a 2 cm cube.

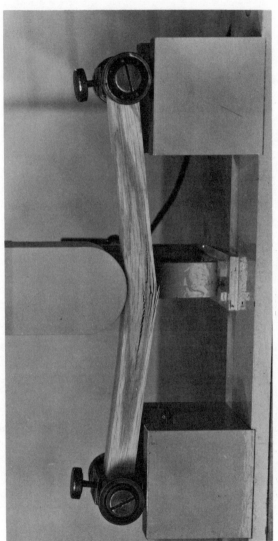

9. Static bending test on 300 × 20 × 20 mm piece showing failure on tension face. C.C.R.

10. Impact bending test on 300 × 20 × 20 mm piece showing failed specimen on left. C.C.R.

11. Compression parallel to grain test on $60 \times 20 \times 20$ mm specimen, showing failure of test piece. C.C.R.

12. Janka hardness test on side surface of static bending test piece with electrical contact to signal depth of penetration. C.C.R.

Tensile strength

The resistance to extension is not easily measured in timber as it is only truly applicable when wood is perfectly straight grained; otherwise failure of the piece in tension will be due to splitting and separating of the fibres.

Effect of Moisture Content on Strength

When all the free water in the cell spaces has disappeared and the cell walls themselves begin to dry out the wood becomes

13. Test to measure shear strength on 20 mm tube. C.C.R.

much stronger in certain respects. In particular the compression strength increases rapidly as the timber dries. Air-dry wood at 12 per cent moisture content (see p. 65 for definition and estimation of m.c.) may carry twice the load unseasoned timber is able to bear (Fig. 15).

It is therefore very necessary to control and measure the moisture content of the test pieces and it is usual in tables showing the strength properties of various timbers to quote figures for the green unseasoned condition and also for air-dried wood with a 12 per cent moisture content.

Not all the strength properties are affected in the same way by changes in moisture content. The toughness, for instance, may even decrease as the wood dries out, so very dry wood,

14. Cleavage test on Monnin-type specimen. Failed test piece on left. C.C.R.

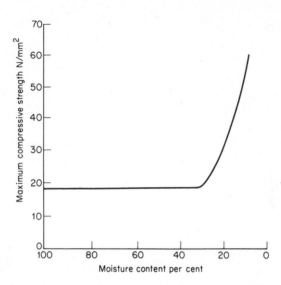

15. Effect of drying on compression strength on pine samples, corrected for specific gravity (based on Figure 3 in *F.P.R.L. Bulletin* No. 50). C.C.R.

though it is stiffer and harder than green timber, may in fact be more brittle.

Effect of Density on Strength

As explained on page 14 the denser a wood is the more wood substance it contains so it is not surprising that there is a fairly close correlation between many of the strength properties of woods and their densities.

Apart from the effect on density of the presence of cell contents such as resins and gums, the specific gravity of a timber is a major factor in determining its strength (see Fig. 16). There is an almost linear relation between maximum compression strength, maximum bending strength, and hardness, and specific gravity. An example is quoted by Lavers who showed that the correlation coefficient between specific gravity and maximum compression strength was 0·815— which indicates a close relation.

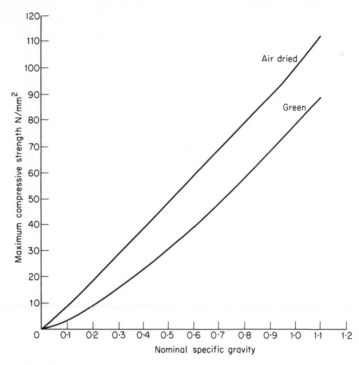

16. Relation between maximum compression strength and nominal specific gravity, based on average of large number of species. (Adapted from Figure 4 of *F.P.R.L. Bulletin* No. 50.) C.C.R.

Effect of Temperature on Strength

It has been shown fairly recently that there is some reduction in the strength of wood with a rise of temperature. Though this is not very significant within the range of temperatures normally experienced it is of interest to note that temperatures well below freezing point considerably increase the maximum compression strength of both hardwoods and softwoods.

Table 1, which is based on Forest Products Laboratory. Bulletin No 50, gives the values in metric units for the various strength properties of five well-known timbers. It will be noted how seasoning green timber down to 12 per cent m.c. increases all the strength values with the exception of the shock resistance. The Table shows that the resistance to compression parallel to the grain and the hardness of the wood are the properties that increase most on drying. The hardness of dry oak is well known to anyone who has tried to drive nails into old oak beams!

It must be emphasised that the values quoted in Table 1 and elsewhere are averages based on a limited range of samples free from defects, and that the weakest and strongest of these samples can occasionally be found to differ so much that the strongest piece could have a strength three times greater than that of the weakest, despite the fact that none of the pieces show any visible defects. When designing a structure it is extravagant to assume that there will be a number of weak pieces, but unless some means is available of identifying them safety will require the designer to use larger pieces than if all are known to have a predetermined minimum strength. This is the reason and basis for stress grading of timber (see Chap. 15).

As mentioned above, all the strength properties that are quoted (as in Table 1) are based on tests on small clear specimens of straight-grained timber free from knots and defects. Today when very little timber of such quality, sawn from large trees felled in virgin forests, is available, commercial supplies of timber include a high proportion of pieces containing knots and other features which affect the strength of the wood. This means in practice that only a proportion of the pieces in a mixed commercial consignment will conform to the average strength figures for clear timber of that particular density.

Growth Features that Influence Strength

The growth features that most seriously affect strength are:

Density, with which rate of growth may be associated.
Knots, which are the bases of side branches embedded by the expansion of the tree.
Deviations of the grain in relation to the line of the trunk.
Reaction wood, which in softwoods takes the form of *compression wood* and in hardwoods is known as *tension wood*.

It has often been the practice to refer to these collectively as defects; but as a high proportion of the commercial supplies of timber, particularly of softwoods, contain some of these features it is unrealistic to claim them all as defects. I prefer to restrict this term to true defects such as insect damage, incipient decay and damage caused by faulty seasoning practices (see Chap. 14).

Density and Rate of Growth

Density has already been discussed on page 43.

Although the rate of growth is often quite closely related to density in many timbers, particularly ring-porous hardwoods, in others there is little correlation.

In hardwoods such as oak and ash, where the earlywood consists mainly of large vessels and few fibres, timber from slow-grown trees contains relatively little of the denser late-wood which consists mostly of thick-walled fibres. So in choosing, for instance, ash for tool handles or sports goods where maximum toughness is required, the wood should contain from 6 to 10 rings per inch. Similarly oak for ladder rungs should possess plenty of fibrous latewood and should come from fairly fast-grown trees.

In softwoods very fast-grown wood tends to have a lower specific gravity than wood from trees that have grown at a moderate rate. Timber with 8 to 20 rings per inch is best for structural purposes.

Wood from the very slow-grown pine and spruce shipped from Archangel is not so strong as that from trees of a moderate rate of growth, and its very fine structure makes it more suitable for joinery than for structural purposes.

Knots

Knots are of two kinds known as 'live knots' and 'dead, or loose, knots'. Live knots are those in which the branches were still living when the wood around them was formed so that the tissue of the branch is intimately connected with that of the trunk. Dead knots on the other hand are those in which the branch was dead when it was embedded in the further growth of the trunk. Such knots have no structural connection with the tissues of the trunk and so are liable to fall out when the the wood is sawn into boards.

Quite obviously the grain of the wood will be displaced from the straight around a knot, the degree of displacement depending on the size of the knot.

The effect of knots on the strength of wood depends on the particular property that one is considering. They have relatively little effect on compressive strength of large posts, and they may actually increase the resistance to cleavage—as anyone knows who has tried to split knotty logs. On the other hand they have a very serious effect on the bending strength of joists and on sawn material such as pallet boards and flooring.

The number, size and position of knots are among the principal features to be taken into consideration when timber is graded on a strength basis. Such grading is discussed in Chapter 15. Methods for measuring the defects and growth features that influence the strength of softwoods to be used for structural purposes are described and illustrated in BS 1860 of 1959.

The deleterious effect of knots on strength may be minimised by cutting the wood into slices and then gluing these together so that the irregularities in the grain are evened out over the thickness of the piece. This process is known as 'lamination' (see p. 176).

Deviations of the Grain

As the cleavage strength of many timbers is quite low stressed members are liable to fail, when loaded in bending, by the fibres splitting from each other along the grain if the angle of the grain deviates appreciably from the long axis of the member. Fig. 17 illustrates a failure of this kind. Many failures in ladder sides are due to the use of timber with sloping grain. Failure of chair legs and rails, and similar small dimension material, is often the result of cross-grain wood having been used. After the fracture the angle of break reveals the sloping grain but this may not be at all obvious when the piece is intact and therefore a special instrument is used to detect it. Many specifications for articles in which strength is critically

17. Failure in ladder stile due to cross grain.

important, such as ladder sides, lay down that the slope of the grain, as determined by this instrument, must not be greater than one in twelve in relation to the long axis of the piece.

Interlocked grain in which alternate concentric zones slope in opposite direction may yield timber that is strong in compression parallel to the grain and very resistant to cleavage.

Reaction Wood

When a tree grows on a slope and there is a pronounced bend in its trunk the texture of the wood may change in the part that is under compression or tension. In coniferous softwoods the reaction wood develops in the lower side of the trunk or branch that is under compression. This so-called *compression wood* is often denser than the surrounding normal wood but it is not correspondingly strong and tough. The chemical composition of the cell walls in it differs from that of normal wood and such tissue shrinks excessively in the longitudinal direction on drying (see p. 14). Compression wood may be recognised by its darker-than-normal colour and the lack of distinction between the early and late zones.

In hardwoods on the other hand the reaction develops on the upper side of the branches and leaning stems and is known as *tension wood*. It is weaker in compression and somewhat stronger in tension than normal wood. Like compression wood it shrinks more in the longitudinal direction on drying than normal wood. It is also paler in colour and its lignin content is lower (see p. 54).

Brashness of Timber and its Causes

Timber is said to be 'brash' when under stress it breaks suddenly and completely with a brittle fracture. Such timber has a low shock-resistance and lacks toughness. Brashness may be due to a variety of causes of which the most important are low density, fungal decay, compression damage, or over-heating.

Low density. Wood that is much below the average density of the species is often abnormally low in toughness. This is particularly the case in the wood from the centre of the tree where the first few annual rings may form a core of 'juvenile wood' much below the rest of the trunk in density and strength.

Fungal decay. Incipient fungal decay that has not reached the stage at which the timber would normally be considered to be decayed or rotten is sometimes called 'dote'. It seriously reduces the toughness of timber even before the density has been significantly reduced.

Compression damage. The structure of wood may be damaged by severe compression stresses suffered either in the growing tree or during felling operations.

The 'brittleheart' that develops in the centre of a number of tropical hardwoods is the result of severe longitudinal stresses imposed on the core of the tree as the tree increases in diameter. Sometimes this results in a visible line across the planed surface of a board and this is known as a 'thundershake'. In less obvious instances microscopic examination may reveal slip planes in the cell walls leading to a compression crease.

Compression damage may also occur in logs that fall across an obstacle on the ground when they are felled.

Any wood showing compression failures is brash and liable to fail if suddenly stressed.

Overheating. Prolonged exposure to temperatures at or only a little above the boiling point of water may result in a serious loss of toughness.

Choice of Timbers on a Strength Basis

When choosing a timber for any purpose where resistance to stress is important it must first be decided what precisely will be the nature of the stresses involved. Below are listed, as examples, some of the major uses of timber together with the strength properties most significant in their utilisation.

Building timbers. (i) For beams, rafters and joists a high bending strength in relation to weight, and adequate stiffness to prevent flexing under stress, are important. (ii) Posts, stanchions and columns require high compressive strength parallel to the grain and, if tall, adequate stiffness. (iii) Flooring boards need to have moderately high bending strength. Flooring that is to be exposed to heavy traffic or heavy industrial use should be very hard and resistant to abrasion, while if it is to be covered with carpet or linoleum moderate hardness will be sufficient.

Poles. Timber for poles must have moderately high resistance to bending stresses.

Marine piling. Very high compressive strength is essential for piling timbers.

Railway sleepers and ties. A high resistance to indentation is required to prevent damage from metal rail fastenings.

Packing cases and pallet timbers. Moderately high bending strength and toughness are required together with the ability to accept nails without splitting—i.e. good resistance to cleavage.

Tool handles, sports goods and aircraft members. For all these purposes it is essential that the wood should be tough and have a high resistance to suddenly applied loads.

BIBLIOGRAPHY

Lavers, J. M. *The Strength Properties of Timbers.* Forest Products Research Bulletin No. 50 (London, H.M.S.O.), 1969.

Silvester, F. A. *Mechanical Properties of Timber.* (Oxford, Pergamon), 1967.

Wangaard, F. F. *The Mechanical Properties of Wood.* (London, Chapman & Hall), 1950.

CHAPTER FIVE

Chemistry of Wood

Composition of Wood

The greater part of the basic wood substance that composes the cell walls of woody tissue consists of cellulose and lignin. Both these materials are polymers; that is they consist of large numbers of similar chemical units bonded together to form very large molecules. Everyone is now familiar with man-made polymers, such as polythene, nylon, terylene, PVC, and so on. In these the individual units have been artificially assembled in such a way as to build up molecules which form materials that have certain specific properties such as toughness or elasticity. It is now known that organic materials which possess high tensile strength have very long molecules consisting of a chain of units bonded together.

In addition to cellulose and lignin wood contains 10 to 20 per cent of other polymers called hemicelluloses. Both true cellulose and the hemicelluloses are compounds known as polysaccharides; that is to say they are polymers built up of sugar units. When these substances are hydrolysed by treatment with an acid—e.g. dilute hydrochloric acid—they break up into their constituent sugars.

Cellulose

Pure cellulose, which has the elemental composition carbon 44·4 per cent, hydrogen 6·2 per cent, and oxygen 49·4 per cent, is composed entirely of glucose units and yields only glucose when hydrolysed according to the equation:

$$(C_6H_{10}O_5)_n + nH_2O = nC_6H_{12}O_6$$

n is a very large number, probably of the order of 10 000.

The formula for β-glucose and the constitution of a chain molecule of cellulose showing the β-linkages are given in Figs. 18 and 19.

18. Structure of molecule of β-glucose.

19. Structure of cellulose molecule.

These long cellulose molecules tend to lie parallel to each other and to form bundles which together form microfibrils. These are individually exceedingly thin, having a diameter of about one millionth of a mm; but when aggregated in regular alignment they form the fibres that can be observed in the electron microscope.

All the important vegetable fibres, such as cotton, flax and ramie, owe their tensile strength to their being composed of cellulose. Other vegetable fibres, such as jute and sisal, contain lignin in addition, but it is to the cellulose that they owe their strength.

X-ray analysis of the fine structure of cellulose has shown that some kinds have a largely crystalline structure though there are usually amorphous regions present as well in the microfibrils. The relative proportions of the crystalline and amorphous regions greatly affect the physical properties of the cellulose.

It will be noted from the formula shown in Fig. 19 that there are a large number of hydroxyl groups along the chain. These adsorb water molecules from any moisture in the air and this leads to a swelling of the cellulose—a fact of great practical importance (see p. 73). Most of the chemical reactions of cellulose occur with these hydroxyl groups. By esterification products such as cellulose acetate and nitrocellulose can be formed.

Hemicelluloses

Hemicelluloses are, like cellulose, composed of large numbers of sugar units, but they differ in that they contain more than one kind of sugar. Some hemicellulose compounds are composed of pentose sugars with five carbon atoms (e.g. xylose); others also contain hexose sugars, such as glucose and mannose, with six carbon atoms. The molecular chains of which they are composed are much shorter than in true cellulose and the chains are frequently branched. In consequence they do not build up fibres but are of a more glutinous nature.

The term 'holocellulose' has been used as an omnibus word to include all the polysaccharide materials in wood, i.e. the combined total of cellulose and hemicelluloses.

The cellulose prepared from wood is of great industrial value as it is the basis of many paper and pulp products (see p. 61).

Lignin

Like the other major constituents of wood lignin is a polymeric material composed of a very large number of units; but its composition has remained in doubt for many years and is not yet fully elucidated. It is an inert, insoluble material that cannot easily be broken down into its component parts, as can the polysaccharides. It now appears that the basic structural unit is the phenylpropane nucleus which is an aromatic benzene ring with a three-carbon side chain (see Fig. 20). These units are joined together by a number of different kinds of linkage so that the lignin molecule is not linear but is a three-dimensional, roughly spherical network.

$$CH_3 - CH_2 - CH_2$$

20. Phenylpropane unit in lignin.

Lignin does not contribute to the toughness of wood but it appears to contribute to its hardness, and as a major constituent of the middle lamella between cells it functions as a cementing material between them. To use a simple analogy one may think of wood as a wet sponge, or a saturated fabric, frozen stiff. The fibres of the sponge or fabric represent the cellulose, and the all-pervading ice which gives the stiffness represents the lignin.

Lignin is the material that characterises the woody tissue of plants and distinguishes them from the soft, unlignified tissues. The word itself comes from the Latin *lignum* which just means wood. So far few commercial uses for lignin have been discovered and large quantities of it remain as a tiresome waste product to be disposed of during the production of cellulose from wood. It does however contribute in no small degree to the formation of humus in soils and represents much of the material left in composts after the polysaccharides have decomposed.

Analysis of Wood Substance

Many kinds of wood have been analysed into their major constituents. An average composition of a number of temperate timbers is given in Table 2. The timbers analysed were thirteen softwood and ten hardwood trees, all from North America.

TABLE 2.

	Softwood %	Hardwood %
Holocellulose	66	76
Cellulose	46	49
Pentosans (hemicelluloses)	8·5	19·5
Lignin	27	21

Two facts emerge from a consideration of these figures. Firstly, hardwoods contain more hemicelluloses than soft woods; secondly, softwoods contain more lignin than temperate hardwoods. It should be noted that many tropical hardwoods on the other hand contain more lignin than do those from the temperate zones.

Extractives

So far we have been considering only the major components of the basic wood substance which is broadly similar in most timbers; but there are in addition smaller amounts of substances which are generally referred to as 'extractives', because they can be extracted from wood by the use of solvents. It is these that give to timbers their characteristic colours and odours, and that confer durability on many species. The best known of these extractives is the resin that is formed by many conifers such as Douglas fir, pines, spruces and larches. Resin is formed in, and moves along, the resin ducts and its production is often stimulated by bark wounds that expose the wood. Its production may also be stimulated as a result of infection of the trunk by diseases caused by fungi. By deliberately incising the bark—known as 'tapping' the trees—the flow of resin can be induced, and large quantities of it are collected in this way from maritime pine trees in Portugal and other countries. The resin is then distilled and the volatile true turpentine is separated from the rosin, both of which find a ready market. The amount of resin varies considerably from tree to tree and the most resinous samples of maritime pine may be 10–15lbs/cu ft heavier than those with less resin.

A few hardwood families also produce resins, notably the

Malayan Dipterocarps, and many Australian Eucalypts produce large amounts of gums called kino.

Chemically the resins are terpenes, which are hydrocarbons
having a structure built up of isoprene units (C_5H_8). The
main component of resin is the diterpene, abietic acid.
In the cypress family (Cupressaceae) compounds called
tropolones occur, which have a seven-membered aromatic
nucleus. The substance thujaplicin, which confers durability
on Western red cedar, is one such compound.

The characteristic colours of the heartwood, which is
formed when the living cells of the sapwood die and air enters
the cells, are due to the presence of various extractives. These
are generally phenolic compounds and include flavanoids,
stilbenes, quinones and lignanes. The tannins are complex
polyphenols which occur chiefly in hardwoods such as oak
and chestnut and in the bark of many trees, as well as in the
fruit and leaves of others such as the tea plant. Tannins react
with, and precipitate, proteins and their use to tan (harden)
leather has been known for centuries. They also play an
important role in conferring resistance to decay in the heartwood of many trees, such as oak (see p. 150).

There are a few tropical timbers that contain poisonous
extractives, such as alkaloids, which regularly cause irritation,
or even illness, in people who come into intimate contact with
them in the unseasoned wood or in its sawdust. The following
woods are reported to have caused irritation under poor
conditions of dust extraction—dahoma, guarea, iroko, makoré,
mansonia, Western red cedar and white peroba. When
working with any of these woods special care should be taken
to ensure that the dust extraction plant is working efficiently.
There are some other timbers, teak for instance, that are
occasionally reported as causing irritation to a few people
who happen to be allergic to some substance which it contains.

Non-cellulose Carbohydrates in Wood

Starch is synthesised in the leaves of trees from the carbon
dioxide in the air by the action of sunlight mediated by
chlorophyll. It is then converted into a soluble form (i.e. sugar)
and transported down to the trunk where it is reconverted

back into the more compact form of starch for storage in the living ray and the parenchyma cells of the sapwood. Nutritive material is thus kept ready for the burst of growth that takes place in the spring.

Some woods, such as oak and ash, have sufficient starch at certain times of the year to nourish the larvae of the Lyctus Powder Post Beetle (see p. 94). Staining the surface of the wood with a dilute solution of iodine will reveal the starch grains to the naked eye.

In spring the sap of certain trees, such as Canadian sugar maple, contains sufficient sugar to warrent collection and concentration of the sap to provide an edible syrup.

Nitrogen Compounds

Only small amounts of protein and other nitrogen compounds are present in wood. These are derived from the residue of the protoplasm in the cells when they were living. Both insects and fungi require appreciable amounts of nitrogen for their growth and the addition of nitrogen to wood increases its susceptibility to these organisms. Therefore if it is desired to accelerate the breakdown of sawdust to form compost, some source of nitrogen, for instance poultry droppings, must be added to it.

Inorganic Substances

All timbers contain a small quantity of mineral matter which remains as ash when the wood is burnt. This is generally less than 1 per cent of the dry weight of the wood though a few tropical timbers contain very much more. The ash consists mostly of salts of calcium, potassium and magnesium such as carbonates, phosphates, sulphates and silicates. Wood ash when supplemented with a source of nitrogen is a valuable fertiliser, especially on soils deficient in potash, and for crops such as fruit trees, tomatoes and potatoes that require a good supply of this substance. In the past, before alkalis were available from inorganic sources, wood ash was used for the manufacture of soap and glass.

In general wood ash contains from 10 to 35 per cent of

potash (as K_2O) but in the ash obtained from burning oak wood from a quarter to nearly half its weight may be potash. The potash content from softwoods tends to be lower than that from hardwoods.

The silica content of most woods grown in the temperate zone is quite low but there are a few tropical woods in which more than half the weight of the ash is silica and it is possible to see the grains of it in the ray cells. Woods with a silica content of over 0·5 per cent of their dry weight cause rapid blunting of saws owing to their abrasive action. Such timbers—Australian turpentine for example—are also very resistant to marine borers, probably for the same reason, that the silica resists the cutting action of the valves of the molluscan borers and, so to speak, 'blunts their teeth'.

Acidity of Wood and its Effect on Metals

The acidity or alkalinity of wood is expressed most conveniently in terms of its pH. The pH scale is a logarithmic measure going from o to 14, 7 representing neutrality. Numbers greater than 7 indicate alkalinity and smaller numbers, acidity. As the scale is logarithmic a single unit change indicates an actual tenfold change, for instance a solution of pH 4 is ten times more acid than one of pH 5.

The pH of most woods lies between 3 and 5 with an appreciable variation between different samples of the same species. At a pH below 4·0 the corrosion of steel is significantly increased and so timbers that are consistently more acid than this may give rise to serious problems under damp conditions. Of the timbers commonly used in this country oak, sweet chestnut, and Western red cedar are the most likely to cause corrosion of metals in contact with them if moisture is present (see p. 141). It is therefore desirable that iron fastenings to be used in contact with such timbers should be protected by sherardising or galvanising. Only galvanised nails should be used to fix Western red cedar siding or shingles.

Some timbers give off volatile acids that may set up corrosion even when there is no actual contact. Unseasoned oak, for instance, gives off acetic acid and it is this which is mainly responsible for its characteristic smell. If metal goods are stored in boxes made of, or containing, unseasoned oak severe corrosion can result. Chestnut wood is another that may have the same effect. Lead, a metal that is resistant to many acids, is susceptible to attack by acetic acid. Damage to lead roofs of churches has occurred when the lead has been laid in contact with unseasoned, or partially seasoned, oak boards or rafters.

Wood that has become permeated with salt, either from prolonged immersion in sea-water or through being treated with salt during seasoning, may also set up corrosion in metals even under relatively dry conditions.

Resistance of Timbers to Chemicals

Some liquids are corrosive to metals and for these vessels made of wood are often used. Timber is also very suitable for the construction of buildings in which chemical processes releasing acidic vapours may be carried out.

Although wood does not resist for long the action of concentrated acids and alkalis some timbers have a very good resistance to dilute solutions. Individual timbers vary considerably in this way. Softwoods are intrinsically more resistant to chemicals than are hardwoods, but some of the dense hardwoods are so impermeable to liquids that they also resist chemical attack quite well.

The following timbers have been used succesfully in this country for vats in chemical plants: Canadian Douglas fir (good quality), pitch pine, peroba de campos, afzelia, greenheart and purpleheart.

Wood damaged by chemicals has a fuzzy look, the individual fibres separating from each other. A useful test for chemical damage is to apply moistened litmus paper to the wood and this will reveal if strongly acid or alkaline conditions have been responsible.

Electrochemical Attack on Boat Timbers

It is, of course, well known that many metals corrode when exposed to sea water, but it is not always realised that this corrosion may cause softening and breakdown of the timber around metal fastenings if the wood is also soaked with sea water. It happens in this way. When two dissimilar metals are in contact with a piece of wet wood containing salt, galvanic action takes place between them and sodium hydroxide is formed at the cathode (the more electropositive metal), while at the anode chlorine is released. This chlorine combines with the anode metal (usually iron) and free hydrochloric acid is formed by further reactions. Copper nails and rivets are frequently used in the wooden planking of boats and the alkali which forms around them as a result of this galvanic action darkens the wood surrounding them and eventually softens it so that the fastenings work loose. This condition is sometimes referred to as 'nail sickness'. African mahogany, which has been used extensively for planking in small boats, is rather susceptible to damage by alkali. Softwoods, such as larch, are much more resistant to this type of damage. Though teak is on the whole resistant to chemical damage the author found localised softening in the teak planking of the *Cutty Sark* where the yellow metal bolts were in proximity to the cast iron frames. However it had taken nearly a hundred years to reach this stage!

The risk of damage of this kind can be greatly reduced by using as far as possible the same, or similar, metals in the vulnerable areas, and by inserting effective electrical insulation between metals in contact with each other and between the metals and the timber. It is particularly important to set deck fittings in bitumen or mastic to exclude sea water from the joints. Any electrical leakages from radio or other equipment should be avoided as small stray electrical currents may greatly accelerate such damage.

Use of Timber as a Raw Material

The ever-increasing demands for paper and board products has led to a tremendous development of the industries that produce wood pulp. Enormous mills, consuming millions of tons of wood annually, have been built in, or near, forests from which logs can easily be transported by land or water.

The technology of wood pulping and paper manufacture is a vast subject to which only the briefest reference is possible in this book. A description of the various processes is given by R. H. Farmer.

The aim of the pulping industry is the separation of the individual fibres of the wood with the minimum damage to their strength. These fibres can then be reassembled in sheets to form paper or cardboard. The two basic ways of effecting this separation are by mechanical grinding or chemical treatment.

Mechanical Pulping

In the ground-wood process the separation of the fibres is achieved by pressing billets against a grindstone while a stream of water is played on them. The water prevents overheating, and carries away the fibres in a porridge-like slurry which passes on to the paper-making machines. There is a high yield of pulp (about 95 per cent) from this process, but it produces a somewhat low-grade, weak paper and is used primarily for the production of newsprint.

Chemical Pulping

By dissolving away the lignin that binds the fibres together their separation can be effected without seriously damaging them. Because the lignin is removed the yield of pulp is reduced to about 50 per cent of the wood, but it can be used to produce paper of a much higher quality.

Two chemical processes are commonly used:

The sulphite process involves cooking woodchips in an acidic solution of about pH 2 containing calcium bisulphite and free sulphur dioxide. Delignification in this process results from sulphonation of the lignin.

The sulphate process uses alkaline solutions containing sodium hydroxide and sodium sulphide. This yields a strong, dark-brown pulp very suitable for wrapping paper, but which can also be bleached to produce other kinds.

A combination of chemical and mechanical processes has been developed in which the wood, in the form of chips, undergoes a mild digestion with sodium sulphite buffered with sodium carbonate or bicarbonate. The softened chips are then defibred in a disc mill. As the product of this method contains part of the lignin and hemicellulose the yield is higher than in the purely chemical process.

Production of Charcoal by Destructive Distillation of Wood

As already mentioned the demand for charcoal for smelting iron ore resulted in the destruction of many forests in the 18th century before it was superseded by the use of coke for smelting, introduced in 1760. Charcoal is still used in case-hardening compounds for metals and as a raw material for making carbon disulphide which is used in the manufacture of viscose rayon. It is a very absorbent material and is used in filters to remove impurities from solutions or gases.

Charcoal was formerly made in pits where small logs were stacked and covered with turves and soil leaving only a small flue down the centre through which the fire was lit. The wood smouldered for five to ten days. Today kilns made of metal or brick are used in which the air supply can be more accurately controlled. The carbonisation of the wood is carried out at temperatures up to 500°C. An exothermic reaction begins at about 270°C and distillation then proceeds without further external source of heat.

About one third of the original dry weight of the wood can be recovered as charcoal. The substances that can be collected by condensing the volatile products of destructive distillation include crude acetic acid (called 'pyroligneous' acid), wood tar, methanol (wood alchol), and acetone. Wood tar has strongly antiseptic properties and makes quite a good preservative though it is somewhat corrosive to metals on account of its high content of tar acids.

Hydrolysis of Wood

As already explained the cellulose of wood can be hydrolysed by acids to produce sugars. Many attempts have been made to utilise wood-waste such as sawdust, of which great quantities are at present burnt, in order to produce useful foodstuffs for man and animals. During the war when Sweden was largely cut off from supplies of sugar many kinds of confectionery were made with sugars prepared from wood. But under normal conditions no process has been found to compete with sugar from sugar beet or cane. A major problem is to obtain sufficiently large quantities of sawdust or waste wood in any one area to keep a large plant operating, and the value of the raw material is so low that it cannot bear the high cost of transporting for long distances.

Though experiments to develop more efficient processes and to manufacture more valuable products from the primary sugars continue in Japan, it is only in the U.S.S.R. that a substantial wood hydrolysis industry is known to exist.

BIBLIOGRAPHY

Browning, B. L. (editor). *The Chemistry of Wood.* (New York, Interscience), 1963.

Farmer, R. H. *Chemistry in the Utilization of Wood*, (Oxford, Pergamon), 1967.

CHAPTER SIX

Moisture in Wood

This is probably the most important chapter in this book because moisture affects the behaviour of wood in a number of very critical ways. Most of the difficulties associated with the use of timber stem from the affinity of wood for moisture and from the shrinkage and swelling that accompany changes in moisture content. Decay occurs only when wood is moist. Paint fails if applied to wet wood. For these reasons a proper understanding of the ways in which the amount of moisture in wood can be controlled and measured is essential if timber is to be used successfully in competition with other materials.

Moisture Content, Definition and Estimation

The moisture content of wood is generally expressed as a percentage of the oven-dry weight of the wood and is calculated according to the simple formula:

$$\frac{\text{wet weight} - \text{dry weight}}{\text{dry weight}} \times 100 = \text{moisture content}$$

The following is a numerical example:

$$\frac{16\cdot4 - 12\cdot2}{12\cdot2} \times 100 = 34\cdot5\%$$

It is thus possible for the moisture content to exceed 100 per cent if the wood is a porous one able to hold more than its own weight in water.

The moisture content of wood is usually determined by drying a sample of convenient size, cut across the grain from the piece under examination, in such a way as to get the average moisture content over the whole thickness of the piece (Fig. 21a). It may be that the surface of the piece, especially if it is a thick one, is much drier than the core. To investigate this, samples should be cut so as to test the core separately from the skin (Fig. 21b). The samples should be weighed and

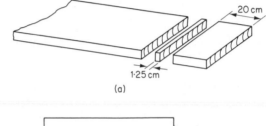

(a)

1·25 cm

20 cm

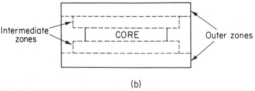

Intermediate zones

CORE

Outer zones

(b)

21. Method of sampling for moisture determination.
 a) Cutting sample from board for determination of moisture content.
 b) Cutting section to determine distribution of moisture.

then placed in an oven at 105°C for a period of about 18 hours—overnight is a convenient period. They are then removed, allowed to cool in a desiccator, and reweighed. Very slight differences in moisture content are of no significance and so it is sufficient to weigh the samples to the nearest centigram.

As we mentioned on page 29 the electrical conductivity of wood is markedly dependent on its moisture content and it is therefore possible to use this property to deduce the moisture content of the wood. But this relation applies only when the content is below the fibre saturation point, i.e the point at which the cell spaces are emptied of water and the cell walls themselves begin to dry out. Many electric moisture meters are now available and are very useful for rapid esti- mation of moisture content, but when using them it is most important to follow the maker's instructions carefully, and also occasionally to check their performance if the readings appear surprising. They generally operate by inserting steel prongs into the wood and since these, of course, can penetrate only a limited distance the readings may not give an accurate estimate of deep-seated moisture. They are also liable to give misleadingly high readings if the surface of the wood has a film of water or ice on it.

A few timbers that contain a great deal of resin, as also those that have been impregnated with creosote, or a similar pre- servative, may lose these volatile substances during oven drying. In this case erroneously high moisture contents would be calculated and therefore an entirely different method is required. In order to determine accurately the amount of water in such material it is necessary to distill it off and separate it from the other volatiles by means of a separating funnel.

Seasoning of Timber

Growing trees contain a very great deal of water. A cubic foot of green oak may contain 22 lb, or over two gallons, of water and, as oak is a dense, impermeable wood, it may take a very long time indeed for this water to dry out if the wood is sawn into thick beams. Practically no drying at all will occur so long as the log remains with its bark on. In fact it will probably begin to decay before any appreciable drying takes place.

It has long been recognised that, except for a few purposes such as marine piling for which whole logs may be used, all timber should be seasoned (i.e. dried). The principal reasons for drying wood before it is used are:

(i) To minimise the amount of shrinkage that will occur when it is used in a structure
(ii) To make it more resistant to fungal decay, staining and insect attack

(iii) To reduce its weight before handling and transport
(iv) To make it suitable for containers and packing cases that will hold goods susceptible to damp and mildew
(v) To render it less corrosive to metal
(vi) To make it suitable to receive paints, varnishes and other finishes
(vii) To render it more absorbent of preservative fluids

Woods vary greatly in the amount of moisture they can hold. This depends largely on the volume of free space in proportion to the amount of cell wall substance—i.e. the density of the wood. In a dense wood there is very little space to hold free water, while in a very light wood, such as balsa, there is so much space that a moisture content of several hundred per cent is possible.

The sapwood of a tree generally has a much higher moisture content than has the heartwood. In some pines, for instance, the moisture content of the sapwood may be over 100 per cent while the heartwood contains 40 per cent or less. There is a great variation also between species. In Douglas fir the average moisture content of the tree when felled is only about 40 per cent. On the other hand, Western hemlock has a very high moisture content even in the centre of the tree.

Air Seasoning

Until about fifty years ago the drying of timber was effected by stacking it in open piles in a well-ventilated place for a long period of time. A rough and ready rule was to allow about a year 'in stick' for each inch of thickness. But however long timber is stacked in the open, even with overhead cover to keep off the rain, its moisture content in this country will never drop much below 18 per cent. The precise figure depends on the kind of wood and also on the humidity of the air around the timber seasoning. If such 'air-dry' timber is then taken into a heated house where the humidity of the air is lower it will lose more moisture and in consequence it will shrink. It used therefore to be good practice to store air-dry timber for a further longish period in a heated workshop before using it for fine joinery or cabinet work.

The basic principle to be observed when drying timber is to remove the moisture from the surface of the sawn timber at approximately the same rate as it is moving from the centre part. If the surface dries out much faster than the centre the surface layers will shrink and split. In practice this means that the flow of dry air over the surface of the timber must be controlled so as not to expose the surface layers to unduly rapid drying. The rate of drying must also be adjusted according to the kind of wood and to the thickness of the boards, planks or scantlings, as well as to the atmospheric conditions prevailing in the drying yard or kiln.

(a)　　　　　　　　　　　　　　　(b)

22. (a) Softwood boards piled to season. Note roof; good foundations; piling sticks between each layer.
　　(b) Oak boards piled in 'boule' to season under cover in shed.

The normal practice is to stack the boards on firm foundations some 18 inches or so off the ground, separating each layer of boards with piling sticks set at intervals along the board. It is important that the sticks should be aligned one above the other in a vertical line so that the load is borne evenly down through the stack, otherwise the boards may bend between the sticks.

Piles should not be more than 6 to 8 feet across so that the air can flow easily through them. They can be of any height so long as they remain stable and tall piles are in fact an advantage as the weight on the lower layers tends to keep the boards flat, space is saved, and one roof covers more timber than if the piles were lower. It has not always been the practice to provide a roof over seasoning piles but it is very well worth while doing so in this country to protect the upper layers from both rain and direct sun.

In general softwoods are much easier to season without serious degradation than are the denser hardwoods which are liable to split and warp if dried too rapidly. Sticks one inch square in section are generally used to separate boards of softwoods, but only half inch thick if the planks are of hardwood, especially if these are piled in hot, dry weather.

It may even be desirable to hang sheets of hessian over the ends of piles of timber such as oak and beech if these are exposed in the green condition to hot sunshine. Metal cleats are sometimes driven into the ends of planks of timbers that are susceptible to splitting.

Kiln Drying

The method known as 'kiln' drying was developed in this country in the 1920s in order to accelerate the seasoning process. In some ways this was an unfortunate word to use as it suggests the idea of baking wood in an oven, and this is the very last thing that ought to be done. 'Controlled' drying would be a happier expression, as in a timber-drying kiln both the temperature and the humidity can be controlled.

A drying kiln consists of a chamber made either with brick or metal walls insulated to prevent loss of heat, and provided with a large door at one end, through which the piles of timber are brought on trolleys. The use of higher temperatures than those usually obtained in air drying is beneficial in two ways: it accelerates the movement of water towards the surface of the pieces, and it increases the rate of evaporation from the surface. Initially the humidity should be kept fairly high and the temperature moderately low. Then as the moisture content of the wood begins to fall, humidity is reduced and the temperature is raised until the wood reaches the desired moisture content. These changes can be made fairly rapidly with easily seasoned timbers in thin pieces, but must be applied over a

23. Maritime pine fencing. Slats piled openly to dry quickly.

matter of weeks when dealing with refractory hardwoods, such as oak.

At the Princes Risborough Laboratory schedules adjusted to the requirements of all the main commercial species of wood have been worked out. For detailed information reference, should be made to their publications dealing with kiln drying which can be dealt with here only in outline.

An important requirement of a good drying kiln is that the conditions should be as uniform as possible throughout the kiln with air flowing evenly through the stack at fairly high speeds, about 6–8 ft per second. This can best be achieved by using large fans, 6–9 ft in diameter, at the side of the stack. Heat is usually supplied by steam pipes and when a boiler is available steam can also be used to humidify the air. However other forms of heating are sometimes used, such as oil-fired air-heaters which are becoming increasingly popular owing to their being comparatively cheap to run. In kilns thus heated humidification can be achieved by disc humidifiers or by small electrically-heated evaporators.

In most kilns the moisture removed from the timber escapes as vapour through vents in the roof. But this involves loss of heat and so in some modern kilns, which have to be completely sealed for the purpose, the moisture is removed by electrically-driven refrigeration dehumidifiers. The condensed moisture is drawn off as water and the latent heat in the steam is given up to the refrigerant and recovered by the circulating air when the refrigerant is compressed and its temperature raised. It is, of course, always important thoroughly to insulate kilns in order to minimise heat losses, and also to make them as vapour tight as possible.

The progress of drying of the timber in a kiln was usually measured by weighing samples taken from the stack which have been placed so as to be easily removed and replaced. In modern kilns the air conditions can usually be controlled automatically by electrically operated instruments. Fully automated kilns have been designed but it is doubtful if their cost can be justified as the requirements of different kinds of wood, and of different consignments of the same timber, vary so much that human judgement is required to achieve the best results.

Advantages of Kiln Drying

Years ago there was much prejudice against kiln-dried timber. Some of this was due to the unfortunate results of the unskilled operation of primitive kilns, but some was merely the instinctive reaction of the 'mumpsimus' who distrusts every new innovation that achieves in hours a task that took his forefathers weeks, or even years, to accomplish. Today, however, kiln drying is accepted as essential for all timbers required for furniture, high-class joinery and flooring.

The main advantages of kiln drying are:

(i) Precise control of the moisture content best suited for the purpose to which the timber is to be put, and the possibility of achieving the lower moisture contents required for timber which is to be installed in centrally heated houses.

(ii) Rapid turnover of stock and reduction in capital requirements necessary for prolonged storage of valuable stock.

(iii) Avoidance of degradation due to fungal staining or insect attack during storage for seasoning.

Combination of Air and Kiln Drying

When dealing with the more refractory hardwoods, such as oak, it is often preferable partially to air-dry the timber and then to complete the drying down to the required moisture content in a kiln. It is much more economical of fuel and power to use the kilns only for the final stage of drying after the moisture content has been considerably reduced by this preliminary treatment.

It is seldom economical to kiln dry timber in thicknesses greater than 75 mm (3 in) and timber of this thickness almost always receives some preliminary air drying.

Hygroscopicity of Wood

Even after wood has been seasoned down to the desired moisture content by air or kiln drying, exposure of the dry wood to a humid atmosphere will cause it again to pick up moisture from the air, and if it is exposed for a sufficiently long time to a saturated atmosphere its moisture content will again approach the fibre saturation point. The rise in moisture content is always accompanied by a greater or lesser degree of swelling. These changes in size, which occur in seasoned timber according to the humidity of the atmosphere, are known as 'movement' of timber.

When wood shrinks or swells the dimensional changes are not the same in all directions. Wood fibres do not shrink appreciably in length when they are dried and so the longitudinal shrinkage of straight-grained wood is negligible. Practically all the shrinkage occurs across the grain, the shrinkage tangentially being greater than the shrinkage radially to the core. The movement values quoted in the descriptions of timbers are generally the amount of movement that the timber undergoes when exposed to a change from 90 per cent to 60 per cent relative humidity, the latter value being that found in normal indoor conditions in England and which, for most timbers, gives a moisture content of about 12–13 per cent. Relative humidity is a measure of the amount of water vapour present in the atmosphere expressed as a percentage of the maximum that could be present at the same temperature. Tangential movements that occur in this range vary from 3 per cent down to 1 per cent and radial from 2 per cent to less than 1 per cent.

Shrinkage from the green condition down to the oven dry condition is, of course, greater and may be as high as 14 per cent in the tangential and 7 per cent in the radial direction. Since there is an appreciable difference in the amount of shrinkage that takes place in a tangential from that which occurs in the radial direction it follows that a piece cut from the green with a square cross section will no longer be square when it has dried, and that varying degrees of distortion will develop during drying depending on the angle of the grain in relation to the long axis of the sawn piece. The thinner the board the greater will be the risk of distortion.

The following terms are used to describe distortion due to seasoning:

Bow = the curvature of a piece of timber in the
   direction of its length.
Cupping = the curvature occurring in the cross-section of a
   piece.

Timbers that 'move' very little and have low shrinkage values should always be used for high-class joinery work. In general softwoods shrink less than hardwoods, yellow pine and Western red cedar being especially good in this respect. There are, however, certain hardwoods, notably teak, iroko, afrormosia and muninga, that have very low shrinkage values and which are therefore very suitable for use in situations where they may be exposed to considerable change in atmospheric conditions.

Timber that has dried under compression is liable to become 'set' and its final dimensions may be less than if it had been dried under stress-free conditions. If the expansion of wood on wetting is largely prevented, as will occur when a barrel made of seasoned timber is filled with beer, the wood will suffer compression, and if it is later dried out the staves will no longer fit together and the barrel will leak. For this reason water butts should never be allowed to dry out completely. During the Second World War many great vats which had always been at least partially filled with wine, were completely emptied and so much shrinkage resulted that they had to be taken to pieces and remade after the war.

Stability and low shrinkage are important qualities to be looked for in timbers for flooring and panelling. Similarly timbers for pattern-making must keep their exact size and shape, and yellow pine has for a long time been the favourite wood for this purpose. Teak is traditionally the timber of choice for ships' decking which may be saturated with water at one time and exposed to tropical sunshine soon afterwards.

The changes in moisture content that follow exposures to atmospheres of different humidities differ a little depending on whether the timber is losing moisture to a dry atmosphere or gaining it from a damp one. For instance in an atmosphere of 60 per cent relative humidity a timber may have 2 per cent more moisture if it is drying down from a wetter condition than if it is picking up moisture from a dryer state. This phenomenon is known as 'hysteresis'.

The equilibrium moisture content of a timber in any given relative humidity tends to be lower at high temperatures than at normal ones, but in ordinary household conditions the difference is not important.

Moisture Content Specifications

Timber should be dried to the right moisture content for the particular purpose for which it is intended. It can be just as wrong to use timber that is too dry as to use it when insufficiently dried. In the first case it will expand when fixed and in the latter it will shrink and open up. It is little use specifying that timber should be 'well seasoned'—a term that used often to appear in older specifications—as timber that is well seasoned for one purpose may be quite wrongly seasoned for another.

The following are the generally recommended moisture contents (per cent) for various constructional purposes in the United Kingdom:

Carcassing timber	22–25
High class carpentry work	18–20
General joinery about	15
Strip and block flooring; panelling in normally heated houses	12
with high degree of central heating	10–11

In the U.S.A., where lower humidities are general, moisture contents not exceeding 10 per cent are advisable for interior joinery. It is well known that many fine pieces of English antique furniture which have stood without serious change for centuries in relatively cool houses in the U.K., suffer quite severely from shrinkage when transported to the very dry atmosphere in centrally heated buildings in the northern states of America. Joints open up and panels split. Damage is likely to be most severe when the change in moisture content occurs suddenly without the timber having a chance to adapt itself gradually to the changed conditions. Central heating should never be turned on at full blast in a new house, but the temperature should be raised slowly over several weeks.

Once timber has been brought to the correct moisture content it must be protected against rewetting either by actual exposure to rain or by prolonged exposure to a damp atmosphere. It is asking for trouble to load manufactured joinery in open trucks and then to dump it uncovered on the grass while building operations are going on. Even without such gross ill-treatment as this, prolonged exposure of high-class joinery to a damp atmosphere will soon raise its moisture content, especially if some surfaces have not received any protective coatings of paint or varnish.

When it is vitally important that no movement should take place, as in fine quality joinery, it should be wrapped in polythene sheeting for transport from the air-conditioned factory in which it has been made, and unwrapped and put in position only after it has been ascertained that the building has fully dried out and that the relative humidity in it has reached its final equilibrium*.

If kiln-dried flooring is laid in a new building in which there is still moisture coming from the bricks and plaster it will swell and may rise up in great mounds of distorted boards. Tests should always be made of the relative humidity and of the moisture content of the walls before high-class flooring and panelling is installed.

From what has already been said it must be apparent that no amount of air drying in the open can achieve the degree of dryness required for high-class joinery to be used in modern centrally-heated conditions.

Treatments and Finishes to Reduce Moisture Uptake

The movement of timber resulting from changes in moisture content is one of its major disadvantages and a great deal of research has been carried out to discover means for preventing or reducing the uptake of moisture by seasoned wood. The

*See *Timberlab News* No 15. December 1972.

treatments are of two types: moisture-resistant coatings, and impregnation treatments.

Moisture-Resistant Coatings

The traditional finish to external woodwork, which consists of a lead-based primer followed by an undercoat and two finishing coats of gloss paint, gives a good protection against the passage of water and greatly delays the entry of moisture in vapour form. So long as this film is maintained in an intact condition the moisture content changes will be much less than those of untreated wood. Nevertheless if damp conditions prevail for a long time the wood will slowly absorb moisture and begin to swell. This is the reason why even properly painted casement and sash windows tend to stick in wet weather. Coatings containing aluminium in leaf form give very good protection against penetration of damp.

A recent development in this field has been the total enclosure of well-dried wooden sills in a plastic covering totally sealed against the entry of damp. Provided this seal remains intact it should give excellent protection, but experience in this country of this type of construction is still rather limited.

Once a moisture-resistant coating is seriously damaged so as to permit water to get into the wood the coating will, of course, retard its evaporation. It has recently been shown that it takes months for painted wood to dry out after it has become saturated with water. Such wetting may occur when a building has been flooded, or when water has penetrated into joinery from adjacent brickwork that has been persistently damp due to overflowing pipes, ineffective gutters and so on.

Impregnation Treatments

Steaming wood reduces its hygroscopicity somewhat, but only to a limited degree. Impregnation with solutions of waxes and resins effectively reduces swelling and shrinking but only over short periods. Probably the most effective treatments are those which involve the impregnation of the wood with synthetic resins that can be polymerised into solids after they have penetrated into the wood. Such treatments greatly increase the strength and hardness of the timber but unfortunately make it very expensive.

Soaking timber in a strong solution of polyethyleneglycol 1000—commonly abbreviated to PEG—greatly improves its dimensional stability. This is a wax-like substance which dissolves readily in warm water and is odourless, non-toxic and non-corrosive. To attain the maximum stability it is necessary to introduce amounts of the substance equal to about 25 per cent of the dry weight of the wood, but lesser amounts are adequate to prevent shrinkage and splitting of wood that is to be used for carving and turnery work (see p. 211). For instance a disc of walnut 9 in across and 2–3 in thick from which a bowl is to be carved will require soaking for 60 days in a 30 per cent solution at 70°F, or for 30 days at 140°F. The wood should be treated when freshly cut but if only dry wood is available the discs should be soaked in water for 2–3 days before treatment.

Soaking softened, water-logged wood from archeological sites in a solution of PEG conserves their shape better than any other known method.

BIBLIOGRAPHY

Bateson, R. G. *Timber Drying and the Behaviour of Seasoned Timber in Use* 3rd edition. (London, Crosby Lockwood), 1952.

Brown, W. H. *An Introduction to the Seasoning of Timber*. (Oxford, Pergamon), 1964.

Mitchell, H. L. *How PEG helps the Hobbyist who Works with Wood*. U.S. Department of Agriculture, 1972.

CHAPTER SEVEN
Enemies of Timber

Timber is a naturally durable material which does not change, or lose its nature, as a result of age. So long as it is protected from moisture and insect attack it can remain unchanged for centuries. Many materials become brittle and crumble after a certain number of years due to oxidation and other slow chemical alterations in their composition, but wood retains its strength indefinitely. Nor does it suffer fatigue, as do metals, after repeated stressing.

Under damp conditions, however, and when in contact with the soil all materials derived from plants and animals decompose, more or less slowly. It is in fact fortunate that they should do so as otherwise forests would become cluttered up with fallen trees, branches and leaves. Decay does not happen as a result of wet conditions alone but is brought about by the action of fungi and, though to a lesser extent, by bacteria.

Fungi are forms of plant life that do not possess the chlorophyll which enables green plants to make organic materials from the carbon dioxide in the air. Consequently they must obtain their nutrients from living on other plants, on animals, or on their dead remains. The fungi that attack living plants are called 'parasites' and those that grow only on dead remains are known as 'saprophytes'.

Some fungi, such as the one that causes Dutch elm disease,

grow in the sapwood and block the vessels that carry the sap to the leaves, thus causing the tree to die of desiccation. Others grow deeper in the trunk causing the disease known as 'heart-rot'; while 'butt-rot' is caused by a fungus entering the tree at its base and may eventually result in the tree falling in a gale.

It is however the saprophytes with which the timber user is most concerned. These fungi live on fallen logs and branches and attack timber in storage, and also wooden structures in service if these are not protected by preservatives or paint.

Some kinds of wood are also attacked by the larvae (grubs) of insects, especially beetles, and it is important to be able to distinguish between fungal decay and insect damage (see below) while timber in the sea may be damaged by marine borers (see p. 100).

The following are the main distinguishing features by which the different kinds of damage may be recognised, but it should be noted that both types of damage may be found in the same piece of timber as once wood is decayed it becomes attractive to many kinds of wood-boring insects.

Fungal decay	*Insect damage*
Wood crumbles into square edged pieces, or into a whitish, lint-like substance.	Narrow tunnels are bored in the wood. These are often filled with fine powder and sometimes contain living larvae.
Toadstools, brackets, or skins of fungal growths appear on the surface.	Small round or oval exit holes can be seen on the surface.
Wood is discoloured, some-times with narrow dark lines through it.	Little discolouration.
There is often a mushroomy or fungal smell.	No noticeable smell.
Occurs in situations that are, or have been, damp.	Occurs in relatively dry situations.

Nature and Life History of Fungi

Fungi can rapidly infect large areas of crops or considerable volumes of sawn timber by means of their minute spores which function as seeds and are produced in fantastic numbers. It has been reliably estimated that a single mushroom 10 cm across can produce 1600 million spores in the few days of its existence; while some of the larger fungi that grow on trees produce hundreds of millions of spores per hour over a period of weeks or months. These spores, which measure only 5–10 microns in length (1 micron $= \frac{1}{1000}$ mm) appear in mass as a very fine powder. They are so light that they can drift in the air for long distances before they fall to the ground. The result is that sooner or later any material exposed to the air, or the soil, will be infected with fungal spores and, if conditions are suitable, these will germinate and begin to grow.

Fungi do not produce a solid tissue built up of cells, as do the higher plants, but they develop fine tubes, called 'hyphae' which branch and form a weft of mycelium (sometimes known as 'spawn'). This mycelium sometimes takes the form of cords or strands by means of which the fungus can spread much in the same way as do the runners of strawberry plants.

Most of the growth of wood-rotting fungi remains hidden within the thickness of the wood so that often the existence of the fungus, and of the decay which it has caused, is not suspected until the fungus breaks out through the surface to produce the fruiting body, or 'sporophore'. These fruiting bodies may be shaped like a bracket or toadstool or may be merely a skin and it is only by the appearance of these that most fungi can be positively identified and named. Once the fruiting bodies have formed they start to produce the spores which reproduce the fungus, but the spores will germinate only if moisture is present, and the fungus will develop only if the material on which the spore has alighted contains the nutrients that it requires.

The bark of trees provides a natural barrier against the entry of spores and hyphae and so wood-rotting fungi can infect the trunk only if there is some wound that exposes the wood. The stumps of broken branches are often the point of entry for the fungi that cause heart-rot in standing trees. When the tree is felled the natural protection of the bark is, of course, lost, and the ends of the logs soon become infected with spores of all kinds of fungi. Some of these may be moulds which stain the wood without decaying it (see p. 192) while others will be the wood-rotting types. The rate at which these can grow into the wood depends on the temperature and on the kind of tree. In all trees where there is a distinction between the sapwood and the heartwood the sapwood is always much more susceptible both to staining and decay.

Fungi, like other plants, grow much faster in warm weather than in cold, spreading about twice as quickly at 70°F as at 50°F. For this reason it is wise to fell susceptible trees, such as beech, during the winter and remove the logs from the woods before the warm weather comes. Under tropical conditions even a few days delay in getting the logs of suceptible species like obeche out of the forest may mean that they become infected with staining fungi and insects (see p. 195). But at temperatures near or below freezing point fungi cannot grow and so in Canada and Scandinavia winter-felled logs do not deteriorate if there is delay in converting them.

Recognition of Decay

Fungal decay of wood is a chemical decomposition brought about by the ferments (enzymes) secreted by the fungi. The wood is in effect digested by the fungus which absorbs into itself certain substances dissolved out of the wood, such as sugars, which it needs for its own growth.

The appearance of decayed wood depends to some extent on the kind of fungus that has caused the rot. Brown rots and white rots are caused by *Basidiomycetes* fungi, while soft rots are caused by micro-fungi.

The main types of decay may be distinguished as follows:
Brown rots. Wood attacked by a brown rot darkens and eventually breaks up into brick-shaped pieces or dark, brittle splinters which can readily be powdered with the fingers. Decay in softwoods is generally of this type.
White rots. In an early stage of a white rot the discolouration may appear as whitish flecks or patches and sometimes there are actual small cavities with a whitish margin. In the more

advanced stage the whole of the wood turns into a white lint-like material, but the decayed wood never disintegrates into powder when handled.

Soft rots. This type of decay generally occurs on the surface of wood that has been exposed for long periods to very wet conditions, as in a water-cooling tower. The wood tends to retain its shape but becomes softened and cheesy. Soft rot is caused by cellulose-destroying micro-fungi which never form large fruiting bodies or skins of mycelium.

The presence of rot in an advanced stage is, of course, obvious, but the early stage—sometimes referred to in the timber trade as 'dote'—may be difficult to detect. Any flecks or streaks of discolouration, be they paler or darker than the normal colour of the wood, call for closer inspection. If decay is suspected the affected area should be tested by inserting the blade of a strong penknife and slowly prising up the grain. In sound wood it should be possible to raise a splinter, but if there is decay the fibres will break off short with a 'brash' fracture. This test can be made more easily in straight-grained softwoods than in hardwoods. The presence of narrow, dark lines like pencil markings is a certain indication of decay in hardwoods. If diagnosis is uncertain, sections of the wood should be examined under the microscope to discover if the fibres of the wood show evidence of fungal attack.

Effects of Decay on the Physical Properties of Wood

The most serious effect of decay in timber is the reduction in strength, which can be quite serious even at an early stage. The first strength property to be affected is the toughness, or shock resistance, which may be significantly reduced even when the appearance of the wood is not appreciably altered, and long before it has reached a crumbling, rotten state. For this reason timber for sports goods and other uses where toughness is critically important, must be inspected with particular care to ensure that every piece is completely sound.

Decay also reduces the density of the wood. If a piece of wood that appears to be normal as to structure and rate of growth is much lighter than usual its soundness must be

24. Brown crumbling rot in softwood caused by the dry-rot fungus, *Merulius lacrymans*. C.C.R.

25. White rot in birch log. Note dark zone lines. C.C.R.

suspected. Severely decayed wood may weigh only a quarter or less of its normal weight.

Decay renders wood more permeable and therefore it absorbs water more readily. If decay has become established in external woodwork the affected areas will rapidly absorb water whenever it rains and conditions will thus become more and more favourable for the fungi that cause the rot.

Decayed wood ignites more easily than sound wood. Fires are sometimes started in a building when a blowlamp is being used to remove paint and the flame strikes a patch of decayed wood.

Prevention of Decay

Felled logs. Logs of species such as beech, poplar and birch, which are particularly susceptible to decay, should be removed from the forest before the end of winter and converted before the warmer weather arrives. If for any reason this is not possible it is wise to coat the ends of the logs and damaged places in the bark with an antiseptic end-coating. An ideal way to store logs that are awaiting conversion is to submerge them in a log pond. This not only protects them against attack by wood-rotting fungi and insects but it prevents the unduly rapid drying of the exposed ends that may lead to splitting. An alternative to total immersion under water is to maintain the logs in a permanently saturated condition by sprinkling them more or less continuously with water.

Sawn timber. Freshly-sawn timber contains a considerable amount of moisture. In some trees water accounts for more than half the weight of the sapwood. It is during the first few weeks of air drying that the timber is most liable to become infected as the surface of the boards is wet enough to bring about the germination of fungal spores. Unseasoned wood must be piled in an open fashion so that the air can circulate over the surfaces of the boards. The piles should be constructed on solid foundations with 18 in clearance between the bottom of the stack and the ground, which must be kept free from vegetation. The accumulation of piles of offcuts and sawdust in a timber yard is undesirable as these may act as breeding grounds for fungi and insects.

Shipment. Sawn timber that is to be shipped overseas should preferably be dried down to a moisture content not exceeding 20–22 per cent. If it is shipped in an unseasoned condition latent infections may develop during the voyage and the timber may arrive in a mouldy, stained condition, or, if the voyage has been long, may even show signs of fungal growth and incipient decay (dote). Much timber is now shipped closely piled in packages. These should be opened on arrival at the timber yard and if the moisture content is found to exceed 20 per cent the boards should be open-piled to further dry out.

When unseasoned timber is to be shipped it is usual today to dip it in an antiseptic solution immediately after it has been sawn up. This gives temporary protection against deterioration during shipment but cannot be relied on to prevent decay and staining for more than a few months.

Timber in Use

The useful life of timber depends on a number of factors including the kind of wood used, the amount of damp and infection to which it is exposed, and the amount of mechanical wear to which it is subjected.

Comparatively little timber is lost through wear compared with the amount that is allowed slowly to decay, so that the term 'durability' has come to mean almost the same thing as 'resistance to decay'. However a timber such as beech, which is not naturally resistant to decay, may be rendered very durable if it is protected by being kept dry or if it is thoroughly treated with a wood preservative.

The relative resistance of a wide range of timbers to decay by wood-rotting fungi has been examined by means of laboratory tests and in field trials in which the life of stakes half-buried in the ground has been followed. As a result of these tests most commercial timbers have been classified into one of five groups.

The durability classification outlined below applies only to the heartwood of the species as in nearly all of them the sapwood is readily attacked by fungi. When the commercial supplies of any timber contain a fairly high proportion of sapwood, as is the case today with European redwood for instance, one must regard the timber as non-durable, though if it were possible to select pieces consisting wholly of heartwood it might qualify as falling into the moderately durable class.

The classes of durability have been defined as follows:

Perishable. Timbers that require rapid extraction and conversion to avoid degradation and stakes of which (2 × 2 in) last 5 years or less in the ground in England.

Non-durable. Timbers which have little resistance to fungal attack under damp conditions and which should receive preservative treatment if used in exposed situations. Stakes of timbers in this class last 5–10 years in the soil.

Moderately durable. Timbers that give good service in exposed situations if protected by paint, but which are not sufficiently durable to be used in contact with the ground. Stakes last 10–15 years in the ground.

Durable. Timbers in this class are suitable for boatbuilding, window sills, gates etc. Stakes last 15–25 years in the ground.

Very durable. Timbers that have a long life even when in situations favourable for decay and are suitable for use, without treatment, as railway sleepers, harbour and dock works, poles and fence posts. Life of stakes is over 25 years.

Examples of timbers in each of these classes are:

Perishable	Non-durable	Moderately durable	Durable	Very durable
Sapwood of most species	Elm	African walnut	Agba	Afrormosia
Beech	Abura	Keruing	Oak	Greenheart
	Whitewood	African mahogany	Idigbo	Opepe
Poplar	Scots pine	Walnut	Red meranti	Teak
Sycamore	Western hemlock	Pitch pine	Sweet chestnut	Kapur
Willow		Larch	Western red cedar	Iroko
		Douglas fir		Yew

It must be emphasised that there are considerable variations in decay resistance within a species and that sometimes samples of a certain timber should be placed in a class above or below that into which its average falls. Even within the heartwood from the same trunk there may be appreciable differences in durability. In African mahogany, for example, the innermost heartwood, formed when the tree was young, is much less durable than the outer zones of the heartwood formed by the mature tree. This is the reason why many old trees become hollow, the innermost heartwood having decayed.

When large quantities of naturally durable wood were available and before modern preservatives were developed, such timbers were used for railway sleepers, poles and fence posts. Today, when many of these woods are scarce and expensive in Europe they are generally reserved for special purposes where their other properties, such as high strength, appearance, and availability in large sizes, are also important, or where the presence of chemical preservatives may be undesirable, as in vats and casks for foodstuffs and wines.

Decay in Buildings

When designing any structure involving the use of timber careful thought should be given to the possibility of moisture reaching and being retained by it. Wherever it is considered that there may be such a risk, either through contact with damp materials such as site concrete, or through leakage (e.g. in flat roofs), or as a result of condensation as in buildings housing swimming pools, then either a naturally durable timber should be chosen or suitable preservative treatment should be given (see Chap. 8). If woodwork decays while the structure of which it is part is still in use it is evident that the wood has not been adequately protected. The extensive substitution of other materials in the place of wood has occurred very largely because it has not been realised that wood can be made completely resistant to destructive agencies. Replacement of wooden windows by metal ones has followed the widespread but unnecessary decay of the former, and the increasing use of Fibreglass for small boats has been due, at least in part, to the inadequate protection of the wood in boats leading to their premature decay.

Dry-rot. This is a term that has been rather loosely applied to the decay of woodwork in buildings and boats, as distinct from the decay of timber in contact with the soil. It is really a rather unfortunate and misleading term for a trouble that is invariably due to the timber having been exposed to persistent damp.

Strictly speaking *dry-rot* in this country should be applied only to decay caused by the dry-rot fungus *Merulius lacrymans*. The specific epithet *'lacrymans'* derives from the 'tears' of moisture which appear on the mycelium when the fungus is growing actively. This particular fungus possesses root-like strands which enable it to spread through brick or masonry walls. The strands contain water-conducting tubes that can transport some of the water necessary for its growth, and so, when an outbreak becomes established in some damp part of a building, the fungus can spread to adjacent areas. When it has been growing for some time—usually about two years—it produces pancake-shaped fruiting bodies which have a wrinkled surface on which are borne the rusty red spores. These are produced in such vast numbers that everything in the area of the fruit bodies may become covered with what looks like red dust.

Eradication of dry-rot requires more drastic treatment than the mere replacement of the decayed timber. A full inspection of the building must be made and every scrap of infected timber removed. Walls infected with strands of the fungus must be sterilized with a fungicide, and all new timber used for replacements, and all sound timber exposed during repair work, must be thoroughly treated with a wood preservative. Most important of all, the source of dampness originally responsible for the outbreak must be sought out and removed. Most dry-rot above ground floor level is due to leakage from faulty gutters and down rain water pipes; while rot at, or below, soil level is often due to rising damp resulting from there being no proper damp-proof courses or to existing ones having been covered over with soil.

Wet-rot. There are a number of other fungi that cause decay of timber in buildings, the commonest of these being the Cellar Fungus *Coniophora cerebella*. The decay which this causes is usually referred to as wet-rot as it does not extend beyond the area which is actually wet. It is more easily eradicated than dry-rot but in the aggregate it causes a great deal of damage.

In recent years decay of external joinery, due to the inclusion of untreated sapwood in window and door frames, and to poor design, has been a major problem, and it is now accepted that all such joinery should receive preservative treatment during manufacture.

Insect Damage

In tropical countries termites (white ants) are the cause of a great deal of damage to timber structures, but in temperate countries damage by insects causes less economic loss than does fungal decay.

Most of the insect damage to timber in temperate countries is caused by the larvae (grubs) of wood-boring beetles that are often spoken of as 'woodworms'. Beetles that cause damage to wood may conveniently be grouped into four classes thus:

Those that attack standing trees—mainly longhorn beetles, (*Cerambycidae*)
Those that attack freshly felled logs—the pinhole borers, (*Scolytidae* and *Platypodidae*)
Those that attack seasoned and partly seasoned logs and sawn timber in storage—mainly powder-post beetles, (*Bostrychidae* and *Lyctidae*)
Those that attack timber in buildings and furniture—furniture beetles, (*Anobiidae*)

Beetles are insects that undergo a complete metamorphosis (change of form) during their life time. The eggs laid by the fertilised female hatch out after a few weeks into larvae and these feed sometimes for years before they pupate (turn into chrysalides) finally to emerge as beetles. Wood-boring beetles are generally rather small, inconspicuous, brown insects, though some of the tropical Longhorn Beetles are quite large and handsome.

Longhorn beetles. With the exception of the house longhorn these beetles are pests of freshly fallen logs in the forest. They usually only attack living trees if these are already in an unhealthy condition.

The larvae, which may be anything from $\frac{1}{2}$ in to 3 in in length, are long, narrow, whitish grubs with a darker head and blackish jaws. They commence boring between the bark and the wood and then bore into the trunk making tunnels which may be up to $\frac{1}{4}$ in or more in diameter, and which are very often oval in section. The length of their larval life depends very much on the suitability of the wood. It may be completed in one year, but if the wood begins to dry out their rate of development will slow down, and sometimes beetles emerge from furniture many years after it has been manufactured. They generally attack hardwoods and the damage is usually confined to the sapwood. The beetles themselves do not feed in the wood and merely bite their way out to emerge as adults.

Damage by longhorn beetles is unlikely to spread once the timber has been sawn up and seasoned and there is little risk of its spreading to adjacent timber. Prevention of damage by these insects largely depends on getting the logs out of the woods as soon as possible after the trees have been felled.

The house longhorn beetle, *Hylotrupes bajulus*, unlike other longhorn beetles, is a pest of seasoned softwoods and has caused widespread damage to buildings in Europe. Though isolated occurrences of the beetle have been reported in different parts of Southern England it is only in one area, around Camberley in Surrey, that it has caused serious damage.

The larvae of this beetle, which are long whitish grubs, markedly segmented, tunnel in the sapwood of softwood beams disintegrating the wood internally while leaving a skin of sound wood on the surface. The exit holes made by the beetles are oval and $\frac{1}{4} - \frac{3}{8}$ in (6–10 mm) across. If an active infestation of this beetle is found it should be eradicated without delay under expert guidance. Only fully treated timber should be used in buildings in areas where infestations have occurred.

Pinhole borers. The pinhole borers or ambrosia beetles are small insects, $\frac{1}{8} - \frac{1}{4}$ in in length, which infest freshly felled logs, often boring deeply into the wood. They are much more common in timbers from tropical and subtropical regions than in European woods. The beetles themselves tunnel in the wood but derive their nourishment, not so much from the wood substance, as from ambrosia fungi (moulds) that grow in the tunnels. These moulds usually cause very dark stains around the tunnels and these are often much more disfiguring than the holes. High-grade veneer logs may then only be fit for inner layers of plywood. No dust is found in the tunnels, which often run across the grain of the wood and rarely exceed 1·5 mm in diameter. There is no risk of the beetles continuing to work in seasoned timber and their attack does not weaken the wood, but as there is considerable prejudice against timber that shows any kind of insect attack, and the public cannot be expected to distinguish between harmless pinhole borers and the more serious infestations, the value of the timber may be seriously reduced. In the tropics it is often necessary to spray the logs immediately after felling with a contact insecticide, but in Europe and the U.S.A. winter felling and early removal of the logs from the woods will do much to reduce the danger of infestation by these beetles of timbers, such as oak, which are susceptible to them. The use of insecticides and how they should be applied is described on page 195.

Powder-post beetles. These beetles, which attack seasoned timbers get their name from the fact that they reduce the wood to a fine, flour-like powder. The Bostrychids are mainly pests of tropical timbers and do not breed in this country, but the Lyctids are cosmopolitan and are significant pests in the U.K. They are the particular concern of the timber merchant as it is wood in the timber yards that is most often infested.

All these beetles infest hardwoods alone and they attack only the sapwood, feeding on the reserves of starch stored in the ray and parenchyma cells. Lyctus beetles are small, narrow, rather stream-lined in shape, one-fifth of an inch long (5 mm) and dark brown in colour. The female deposits her long thin eggs within the vessels of the wood and it is therefore very rare to find larvae in timbers in which the diameter of the vessels is less than that of the ovipositor. Small-pored woods, such as beech and poplar, are seldom, if ever, attacked. The larvae are curved, with a whitish body and yellowish head, dark brown jaws and about $\frac{1}{4}$ in (6 mm) in length. The exit holes are small—about one-sixteenth of an inch across—and they may appear on the heartwood face of flooring although the beetles feed only in the sapwood. Piles of very fine impalpable dust may sometimes be seen on the surface of infested timber, but often the first sign of infestation is the emergence of the beetles and the appearance of their exit holes. If evidence of wood-boring insects becomes apparent within a few months after the purchase of furniture or the installation of hardwood floors or panelling, it is usually due to lyctus beetles having infested the timber before the goods were manufactured.

Prevention of damage by powder-post beetles depends to a large extent on good hygiene in the timber yards and saw mills. All offcuts and scraps of susceptible hardwoods should be removed, and stocks of susceptible timbers should be inspected at least twice a year. If attack is found in an early stage the infested wood can be sterilised by heat treatment in a kiln at about 55°C in a damp atmosphere for about a day.

For detailed recommendations of schedules suitable for different timbers and different thicknesses reference should be made to F.P.R.L. leaflet No. 13, *The Kiln Sterilization of Lyctus infested Timber*. Although this treatment kills any insects that are in the timber it does not prevent reinfestation and if wood has subsequently to be stored during the summer it should be thoroughly sprayed with 0·5 per cent emulsion of gamma B.H.C. or dieldrin.

Furniture beetles. Furniture beetles are mainly the concern of the house-owner for the damage they cause is in buildings and their contents. A great deal of publicity has been given to the need for remedial treatments to eradicate these insects and a thriving industry has grown up to carry out *in situ* treatment of infested timbers, particularly in roofs. Infestations increase but slowly over the years and a house or piece of furniture may be twenty years old before the damage becomes obvious. Bletchly, in a recent Government publication, states that 'failure to recognise this fact has frequently led to undue alarm and excessive remedial measures'.

Two species of beetle in this family are responsible for most of the damage to timbers in the United Kingdom. By far the commonest of these is the common furniture beetle— *Anobium punctatum*. The other is the death-watch beetle— *Xestobium rufo-villosum* which is rarely found in softwoods but attacks large oak timbers, generally those in which there is already incipient decay.

Common furniture beetle. This beetle in a state of nature breeds in dead branches of trees and shrubs. It infests both hardwoods and softwoods, but normally damages only the sapwood in those timbers, such as pine and oak, where there is a distinct heartwood. Many tropical timbers (though not mahogany) and most types of chipboard and hardboard, appear to be immune to its attack. Older types of plywood made with a protein glue (e.g. blood, casein or animal glue) are highly susceptible as the protein content of the glue greatly assists the nutrition of larvae in wood that is naturally deficient in nitrogenous compounds. Wickerwork is also a favourite material of the furniture beetle, and old wicker baskets and trunks, plywood tea chests, etc, should never be stored away in attics as it appears that the infestation of roof timbers is most commonly due to the introduction of articles already infested with the larvae or eggs of the beetle.

26. Damage in a floor board caused by larvae of common furniture beetle, *Anobium punctatum.* × ¾.

Attack by 'woodworm' can be detected by the appearance of small round exit holes, about one-sixteenth of an inch (1·6 mm) across, along the edges of rafters or joists and in the sapwood portions of floor boards, or in pieces of furniture. But if these holes look old and dirty and are full of dust or paint the attack is probably an old one and there may be no living larvae left in the wood. Recent activity may be assumed if the holes are clean and if there are small piles of bore dust ('frass') around them indicating that the larvae are working within the wood. Under a lens this bore dust is seen to consist of tiny lemon-shaped pellets.

The adult beetles, which emerge in June and July, are brownish-black and about one-tenth to one-fifth in (2·5–5·1 mm) long. They can fly actively especially in hot weather and may be found on window panes in infested rooms.

Prevention of attack depends on:

Treatment of susceptible sapwood with a wood preservative —boron compounds are both effective and cheap (see p. 111). Removal of sources of infection—e.g. old furniture in roof spaces.

Regular polishing of floors and furniture. Eggs are always laid on rough, unpolished surfaces such as are often found at the back of pieces of furniture, and particular attention should be paid to the treatment of such surfaces. A coat of varnish that blocks up the pores and cracks in the wood is a useful deterrent.

Use of insecticide vapours and smokes. Vapona strips impregnated with dichlorvos give off a vapour that is fatal to beetles in the air, and if the roof is boarded, hanging up these strips in the roof space every summer will prevent infestation.

If infestation has taken place and remedial treatment is thought necessary a careful survey should first be made to determine how extensive is the infestation. Dust should be removed from the surface of boards, and if the edges of rafters are severely damaged the softened portions should be cut away with a draw knife before applying an insecticide. Many proprietary insecticides are now available and one of these should be applied liberally by brush or a gentle coarse sprayer to all infested timber. The manufacturers recommendations for use should be followed implicitly, and great care should be taken to avoid the risk of fire when applying insecticidal solutions that contain flammable solvents. When dealing with small areas such as infected pieces of furniture the squirting of an insecticide into each individual hole will usually prove satisfactory.

Death-watch beetle. This beetle has achieved notoriety partly because of its name and the superstition that its tapping presaged a death in the house where it was heard. In actual fact the tapping is a mating call and is made by the adults hitting their heads against the wood. It has also become notorious because it has caused damage to the timbers in many famous historic buildings, for example, Wesminster Hall. It is practically unknown in Scotland and very rare in Ireland, occurring most frequently in the southern half of England where in nature it breeds in decayed parts of old oak and willow trees.

27. Damage in oak beam caused by death-watch beetle, *Xestobium rufo-villosum* × ½.

The beetle is larger than the common furniture beetle and makes exit holes about ⅛ in (3 mm) across. It tunnels deeply into large timbers usually spreading from the partially de-cayed ends of beams embedded in damp walls. The beetles are very sluggish and seldom fly, except at temperatures that are rarely reached in large stone buildings. Infestations have in many cases probably been due to the introduction of old timbers already containing the larvae.

Prevention of attack depends to a large extent on protection of the ends of oak beams against dampness and decay. Eradication is often difficult because the larvae bore into large timbers to considerable depths and surface application of insecticides does not always result in sufficient penetration to ensure killing them. Injection of the fluids through holes made in the beams may help to achieve deeper penetration. Strengthening of damaged timbers may be required and further entry of damp must be prevented. Repeated fumigation of roof spaces each summer with insecticidal smokes is an inexpensive way to reduce the beetle population and to prevent reinfestation. In H.M.S. *Victory* fumigation of the whole ship with methyl bromide gas effectively reduced the population of beetles that were breeding within the thickness of the hull. It is wise to seek expert advice when dealing with an extensive infestation of these creatures as structural repairs are often required in addition to insecticidal treatment.

Termites. Termites occur throughout the tropics where they cause an immense amount of damage to wooden structures. They are also quite common in many warm temperate countries, while a few hardy species survive in parts of France, Italy and Spain.

All termites are social insects living in colonies as do the true ants but they belong to a quite different and unrelated order. There are many different kinds which differ greatly in their habits and destructiveness. From a practical point of view they can be divided into two main groups, the earth-dwelling termites, which always maintain a connection with the soil; and wood-dwelling termites, which spend their lives in wood.

All termites avoid the light and many construct tunnels with particles of soil stuck together so that they may pass over inert materials on their way to their food supply without being exposed to their enemies. Working inside the wood, the damage they cause often remains undetected until the part of the structure that is infested receives a sudden knock and collapses.

A termite colony consists of one or more fully developed females, or 'queens' and a few males, along with large numbers of sterile individuals of two types known as 'workers' and 'soldiers'. The queen may live for many years producing eggs at an astounding rate, estimated to reach up to 30 000 per day; so the reproductive powers of the colony are fantastically high.

There are a number of timbers that are naturally resistant to termite attack including the following: afzelia, afrormosia, iroko, muninga, jarrah, greenheart, Rhodesian teak, teak and wallaba. Impregnation of permeable woods with creosote or copper chrome arsenic preservatives gives good protection to susceptible timbers. It is beyond the scope of this book to deal with this subject more fully and for further information about these insects reference should be made to some of the publications listed at the end of this chapter.

Marine borers. Damage to timber in the sea is caused by creatures known as marine borers, and is more severe in tropical and subtropical waters than in the colder waters around Great Britain. Two types of animal are involved—molluscs, which are commonly known as 'shipworms' owing to their somewhat wormlike bodies, and crustaceans, called gribble, which look like wood-lice.

Molluscs—shipworm. The shipworms in the sea around Great Britain are species of *Teredo*. The young free-swimming larvae enter wood through mere pin-holes, but the tunnels they make soon become wider as they grow and bore deeper into the wood and in time they may become nearly an inch (2·5 cm) across. These tunnels are lined with a calcareous deposit and they never contain bore dust. Occasionally timber that has been cut from logs which have been floating for some time in sea water contains old shipworm tunnels and the damage may be taken for insect attack. Its true origin can, however, be deduced by noting the mineral lining in the tunnels.

In warm weather severe damage to susceptible timbers can be caused by these creatures in a comparatively short time. Timbers that are resistant to their attack include the following: afrormosia, basralocus, greenheart, jarrah, opepe, and Australian turpentine. It may however be cheaper to protect a susceptible timber with a sheathing of metal or nylon, or to impregnate it with a heavy loading of creosote. Shipworms in a boat will die after a few weeks if the boat is removed from salt water to fresh water.

Crustaceans—gribble. The *Limnoria* species that attack wood in the sea are small representatives of the lobster and shrimp order. They are widespread in the seas around Great Britain and can grow at lower temperatures than the shipworms. They attack the wood on its surface, making many shorter and narrower tunnels than those made by *Teredo*. The tunnelled surface wears away with the action of the waves and the affected timbers assume a waisted appearance that is typical of gribble damage. They cannot survive in water containing less than 1·0–1·5 per cent salinity.

The timbers that resist *Teredo* are also resistant to gribble, but it is less easy to protect wood against them with chemical preservatives. A physical barrier, such as a good layer of paint, is a fairly effective protection for boats. Heavy concentrations of copper naphthenate or of copper chrome arsenic preparations are recommended for protecting susceptible woods against this type of damage.

BIBLIOGRAPHY

Bletchly, J. D. *Insect and Marine Borer Damage to Timber and Woodwork.* Forest Products Research Laboratory (London, H.M.S.O.), 1967.

Cartwright, K. St. G. and Findlay, W.P.K. *Dry-Rot in Wood.* Forest Products Research Bulletin No. 1 6th edition. (London, H.M.S.O.), 1960.

Cartwright, K. St. G. and Findlay, W. P. K. *Decay of Timber and its Prevention* 2nd edition. (London, H.M.S.O.), 1958.

Findlay, W. P. K. *Timber Pests and Diseases.* (Oxford, Pergamon), 1967.

Gareth Jones, E. B. and Eltringham, S. K. (editors) *Marine Borers, Fungi and Fouling Organisms of Wood.* O.E.C.D., 1968.

Harris, W. V. *Termites: Their Recognition and Control* 2nd edition. (London, Longman), 1971.

CHAPTER EIGHT

Preservation of Timber

Generally speaking the term 'wood preservation' implies treatment of timber with certain chemicals that are poisonous to the fungi and insects that attack it, but surface deterioration can be prevented by physical means such as paints and other finishes. It must, however. never be forgotten that the best way to preserve timber is to keep it dry—i.e. at a moisture content below 20 per cent. No wood-rotting fungi can attack wood that is dryer than this, while under very dry conditions, below 10–12 per cent, even the most hardy insects cannot breed. Wooden objects thousands of years old have been recovered from Egyptian tombs because the extreme dryness of the climate prevented biological deterioration.

This chapter describes the various finishes that can be given to wood, the main types of wood preservatives now in use and the means by which they can most effectively be applied.

Physical Protection of Wood

Destructive agencies can be prevented from reaching timber by a physical barrier which, so long as it remains intact, prevents either fungi or insects from gaining access. Paint, of course, is one preservative of this kind. But timber left in the natural state without any surface treatment undergoes deterioration through other causes than fungal and insect attack. Timber out of doors undergoes a process known as 'weathering'; while indoors the colour of timber changes under the influence of light and its surface becomes soiled with dust and dirt which penetrate the open pores.

External Woodwork

The weathering of external woodwork is brought about by a variety of adverse influences. Exposure to rain washes away the natural colouring matter in the wood, while alternate wetting and drying sets up stresses that lead to checking and splitting of the surface. Frost causes expansion of the water in the cells and this tends to break up the surface of permeable woods; wind carrying particles of sand can quite quickly 'sand blast' wood, eating away the softer layers. Ultra-violet radiation causes the breakdown of the cellulose on the outside of the wood with darkening of the surface; and under humid conditions micro-fungi cause slow soft rot.

The amount of surface weathering these influences can cause depends on the nature and density of the wood. Some naturally oily woods that are not prone to splitting, such as teak for example, suffer very little, but others that 'move' a great deal with changes in moisture content are liable to develop deep cracks into which damp and the spores of wood-rotting fungi may find their way, setting up internal decay.

Protective finishes

Exterior woodwork can be protected against weathering and discolouration and other destructive agencies either by coating with paint, or by applying a transparent preservative which does not obscure the grain. Formerly most external woodwork was painted but the high cost of the labour involved has made reappraisal of other methods of protection desirable. At the same time the decorative qualities of hardwoods—that paint would obscure—are being increasingly appreciated in cladding and external joinery. There has consequently been much research into alternative treatments that will reduce annual maintenance costs—which in the U.K. in 1972 amounted to approximately £85 million—and also retain the attractive appearance.

Paint. The use of paint is a well understood process and, provided that the wood is sound and dry when it is applied and the paint film is renewed regularly, it gives very good protection to external woodwork that is not in contact with the ground. If properly applied it should last without renewal for at least six years. Unfortunately, however, paint films are not impervious to moisture in vapour form, and movement of the wood with seasonal changes in moisture content often leads to the film cracking and allowing entry of decay. Also the adhesion of paint films to some hardwoods is sometimes poor. Oak window sills, for instance, seldom retain paint well and are better preserved by annual applications of linseed oil. Then again painting is often carried out under unsuitable weather conditions and in consequence the film does not adhere firmly to the wood.

Woodwork should be clean and dry when it is painted and exterior work should be treated with a white lead, oil-based primer conforming to BS 2521 before subsequent coats of paint are applied. The National House Building Registration Council, and many Building Regulations, also require that softwood cladding (other than Western red cedar and sequoia) be treated with an approved wood preservative before it is painted.

The hulls of wooden boats are usually protected by regular painting which gives protection against gribble attack as well as preventing the entry of water. It may, however, be an advantage not to paint the inside of the hull but rather to treat it with a solvent type wood preservative. This will permit any moisture that had found its way into the timber of the hull to escape if the bilges are well ventilated.

Plastic coatings. A recent development of an impervious coating as a protection to wood has been the encasing of external joinery timbers in a thin plastic sheathing that is completely sealed. This enables the strength of the timber to be combined with the durability of a plastic finish, and it does not require periodical redecoration as does wood with an ordinary paint finish. The ultimate durability of this kind of material has yet to be determined by practical trials, but it is expected to be good.

Tar oils. The cheapest treatment for external woodwork is a brush-applied coating of creosote, or of a refined tar oil preservative, which can be renewed every 3–5 years; but this does not entirely prevent surface splitting and checking. Impregnation under pressure with creosote gives good permanent protection that requires no subsequent maintenance and is ideal for gates, fencing and farm buildings (see under Chemical Treatments, p. 109). But this method leaves the wood in rather a sticky, smelly condition which renders it unsuitable for some domestic uses especially as creosote sometimes 'bleeds' and stains plaster or mortar with which it is in contact.

Clear Finishes

The increasing interest in using natural wood as a decorative feature of building has focussed attention on the use of clear, transparent finishes to preserve and enhance the natural appearance of the wood. These finishes are of two types, varnishes and water-repellent solutions.

Varnishes. There is no doubt that the attractive appearance of external woodwork is best displayed by applying a clear varnish. Unfortunately no varnish has yet been discovered that will last for more than a year or two without deteriorating so much that it becomes detached in places. Discolouration of the wood thus exposed soon follows. At least three coats of any varnish is required to give a film that will last for two years, and after this regular revarnishing every two years is necessary to maintain a good finish. This involves washing down the surface, removal of loose and flaking old varnish and sanding down to a firm edge. Bleached areas may then have to be stained to bring their colour back to that of the original wood. Altogether maintenance costs will be high.

Most varnishes are based either on alkyd resins or oleoresins. Polyurethanes have not in general given satisfactory results as exterior finishes. The Timber Research and Development Association has published a very useful list of the 62 proprietary varnishes that proved to be the most satisfactory of 218 that were tested by the Association.

Water-repellent finishes. These are liquids that penetrate the surface of wood, allowing the natural pattern of the grain to remain visible, but not forming a glossy surface film. The traditional material for this type of finish was linseed oil, but this tends to hold the dirt and surfaces treated with it may darken so much that they become almost black. Modern water-repellent finishes consist of white spirit solvent containing waxes and resins to give water repellency and a pigment to enhance, or restore, the colour of the wood. In addition they often contain a fungicide to prevent the establishment of wood-rotting fungi and staining moulds. These products give longer protection than varnishes, and retreatment, which may be required after three or four years, is much easier as it does not involve removal of any damaged film. The better products of this type effectively protect wood against weathering as they greatly reduce the uptake of water during rain and the pigments in them absorb some of the light that is destructive to clear varnishes. The use of these finishes is expected to increase substantially as their advantages become better known.

Internal Woodwork

Woodwork that is left untreated in any way tends under the action of light and air either to fade or, in the case of very light-coloured woods, to darken. This means that practically all woods tend to end up a similar dull pale-brown colour, and light-coloured woods left with the grain exposed are also likely to become soiled by the deposition of dust from the atmosphere. For these reasons it is wise to treat all exposed timbers with some form of protective coating. This will also help to protect them against possible infestation by woodworm.

The wearing quality of floors made of any but the hardest woods benefits by the provision of a hard superficial film so that the surface of the wood is not scuffed away. Very hard wearing surfaces are provided by solutions of polyurethanes, urea formaldehyde resins and epoxy resins. Many effective proprietary brands of floor seals are now available and they are easy to apply. If the floor is to be stained the coloured solution should be applied to the clean wood and allowed to soak in and dry before a varnish finish is applied. If a coloured varnish is applied directly it will wear away to disclose unstained wood.

Chemical Wood Preservatives

Historical

The first really effective wood preservative to be used on a large scale was coal-tar creosote. For over a hundred years this was the accepted standard preservative for external woodwork.

The first water-soluble wood preservative was mercuric chloride. This was applied by a steeping process called kyanisation, but owing to the intensely poisonous nature of this chemical and to its corrosive action, its use was always restricted. The use of copper sulphate as a preservative for wood was patented by Boucherie in 1838 who recommended that it should be applied by a sap replacement method (see p. 120). During the nineteenth century this treatment was widely used in Europe and especially in France.

Also in 1838, the use of zinc chloride solutions applied by impregnation was patented by Burnett ('burnettising').

At the beginning of this century Malenkovic and Wolman introduced the use of fluorides for the preservation of wood. These salts, which are toxic to both fungi and insects, are very soluble and therefore readily leached out of the wood. Many attempts have been made to increase their permanence for external use. The addition of chromates was found to fix various toxic metals including copper, arsenic and other elements. The German patent literature is full of references to various combinations of fluorides, dinitrophenol and chromates, sometimes with the addition of arsenic for the purpose of increasing the toxicity of the mixture to insects. Fluorides are still widely used in Germany for the preservation of building timbers but in Great Britain they have never been as popular as they were in Central Europe.

Modern Wood Preservatives

There is no one wood preservative ideal for all purposes as wood in different situations requires varying degrees of protection depending on the hazards to which it will be exposed.

The following are the requirements of any good wood preservative:

The product must be sufficiently toxic at a convenient concentration to render the wood immune to the attack of the organisms to which the treated wood is likely to be exposed.

The preservative must be capable of penetrating a permeable timber to a considerable depth when applied by pressure impregnation (so, for example, it must not be unduly viscous at the treating temperature).

The preservative must persist in the treated wood for many years in an active form. This is a requirement that is not necessary for most pest control products used in agriculture.

It must not be corrosive to metal, or cause deterioration to the wood itself. This in practice means that no product that is strongly acidic or alkaline in reaction can safely be used.

The product must not endanger the health of operatives applying it, nor should it render the wood poisonous to those handling it, or coming into contact with it in service. It must not increase significantly the flammability of the wood in service. Preferably it should reduce this. It must be reasonably inexpensive for bulk users.

Special uses may call for additional requirements. For instance substances to be used on building timbers must not give off strong, persistent odours. Preservatives for timber that will come into proximity with foodstuffs must never give off any vapour that may cause tainting, and those to be used on joinery should have no effect on paint which is applied subsequently.

Commercial wood preservatives available today fall into the following classes: the tar oils; water-borne preservatives, which are either highly fixed or leachable after application; organic solvent types; and emulsions of organic preservatives.

Tar Oils

Creosote distilled from coal tar has long been the principal preservative oil. It is a complex mixture of organic compounds with a distillation range of approximately 200–400°C.

Creosote contains:

Tar acids such as phenol, cresol and higher homologues.
Tar bases such as pyridine.
Neutral components such as naphthalene, anthracene, and other neutral hydrocarbons.

There has been much debate as to the relative importance of these constituents. Formerly it was thought that much of the toxicity was due to the tar acids, but it has now been shown that removal of these does not significantly reduce the effectiveness of the creosote. There has also been much discussion as to what proportions of the creosote should consist of higher-boiling fractions. For permanence in the wood it is necessary to include a fairly high proportion of these.

The heavier oils are of course more viscous and must be heated to secure proper penetration. It is therefore desirable to use this type in pressure-treating plants where the oil is heated to a temperature not exceeding 99°C, and to use the lighter oils only for surface application by brushing or steeping.

The British Standard specification (BS 144 of 1973) stipulates that creosote for general use in treatments involving heating the oil, should have the following composition:

Not more than   6% shall distil at 205°C
Not more than 40% shall distil at 230°C
Not more than 78% shall distil at 315°C
Not less than 60% shall distil at   355°C

The composition of creosote designed for use at ambient temperatures is covered in BS 3151.

There are a number of proprietary tar-oil preservatives prepared from creosote which are designed for surface application by brushing, spraying, or dipping. These are supplied in various colours and they are more suitable for indoor use than is ordinary creosote.

Creosote is eminently suitable for the preservation of railway sleepers, telephone and electric transmission poles, marine piling and fence posts. Tar-oils are not suitable for treatment of timber that is likely to come into contact with food as its smell can cause tainting. They should not be used on timber that is subsequently going to be painted, nor in greenhouses since the fumes are harmful to some plants.

Water-Borne Preservatives

Of the many hundreds of water-soluble mixtures which have been suggested as wood preservatives only two are in general use today in Great Britain.

Highly-fixed water-borne preservatives. These preservatives owe their permanence to a reaction that takes place between their components after contact with the cell walls of the wood, causing the active ingredients to become converted into insoluble compounds which are highly resistant to leaching. This co-precipitation is brought about by the addition of chromates.

One of the first succesful preservatives of this kind was Celcure, patented in 1926 by Gunn in Scotland. This contained copper sulphate together with a chromate. The effectiveness of copper chrome preservatives was later found, by Falck and Kamesam, to be enhanced by the addition of arsenic which greatly increased their toxicity to insects.

Slightly varying proportions of the principal ingredients, copper, chromate and arsenic, are used by different manufacturers, but the composition of the leading brands is approximately:

Copper (calculated as $CuSO_4.5H_2O$)	33 parts
Dichromates (calculated as $K_2Cr_2O_7$)	40 parts
Arsenic (calculated as $As_2O_5.2H_2O$)	20 parts

Impregnation with a solution of this type gives long lasting protection to timber even in situations such as water-cooling towers where the treated timber is exposed to continuous leaching.

The following Table gives the suggested loading of these salts that is required for protection of timber in different situations. The loading is calculated from the weight of solution absorbed per unit volume of wood and the concentration of solution used.

Interior timbers	structural timbers, joists, floor boards	0·25 lb/cu ft
Exterior timbers	window frames, fencing, decking.	0·33 lb/cu ft
Marine work	piling, jetties, etc.	0·5 lb/cu ft
Cooling towers	structural timbers.	0·33 lb/cu ft
	slats in stack.	1·0 lb/cu ft

Leachable water-borne preservatives. For internal use, where the timber is not in contact with the soil or exposed to leaching, treatment with a water-borne preservative that does not become 'fixed' in the wood can give adequate protection against insect attack and decay induced by damp conditions. Preservatives based on fluorides and bifluorides are still used for this purpose in Europe, particularly in Germany, but in the United Kingdom and in New Zealand, boron compounds (particularly sodium octaborate) are preferred. These are generally applied by the diffusion process (see p. 118) to give loadings of 0·2 per cent HBO_3.

While boron compounds are toxic to wood-destroying organisms they do not present any health hazards to man or to domestic animals, and their cost is lower than most other water-borne preservatives. For the treatment of dry-rot in buildings a 5 per cent solution of sodium pentachlorophenate is generally used.

The following are the principal advantages of water-borne preservatives:

Clean condition of timber after treatment with no subsequent 'bleeding' of oil in hot weather.

Timber can be painted after treatment—once the water has dried out.

Transportation charges on the preservative are low as it is shipped in concentrated, solid form.

They do not present a fire hazard—as may happen when preservatives containing flammable solvents are used. They can, in fact, be conveniently combined with fire retardant salts.

They cannot diffuse into, and cause staining in, plaster with which the treated wood may come in contact.

They are mostly odourless and may therefore be used in buildings where foodstuffs are stored.

They are well-suited for sterilization of damp brickwork which is infested with strands of dry-rot fungus.

The disadvantages of the water-borne preservatives when applied to seasoned timber are:

The water used as a solvent to introduce the preservative must afterwards be removed, either by kiln- or air-drying, before the treated wood can be used.

The introduction of water into wood that has been cut to precise sizes causes swelling and sometimes irreversible distortion.

These disadvantages, however, do not arise if the preservative is applied to green timber by the diffusion treatment and the seasoning carried out afterwards (see p. 118).

Organic Solvent Preservatives

Compared with the tar oils and the water-borne preservatives the organic solvent types are a recent development. Basically they consist of a substance that is toxic to fungi and insects dissolved in an organic solvent. This is most commonly a petroleum distillate. When the solvent evaporates the active ingredient left in the wood is resistant to leaching by water. The substances now used in the United Kingdom for this type of preservative include copper or zinc naphthenates, pentachlorophenol, and tri-n-butyltin oxide (subsequently referred to as Tn/BTO). Gamma-benzene hexachloride (B.H.C.) is often added to increase the toxicity of the preservative to insects.

Generally an anti-blooming agent is included to prevent concentration and crystallisation of the dissolved substance on the surface of the treated wood when the solvent evaporates.

There are many proprietary formulations containing varying proportions of these insecticides and fungicides, sometimes in colourless and sometimes in coloured solutions. The solvent chosen may be slow drying or a lower-boiling fraction that is quick drying.

The advantages of organic solvent preservatives are:

They cause no swelling or distortion of seasoned wood. This makes them particularly suitable for joinery.
Once the solvent has dried off the treated wood can be painted.
They penetrate readily and are therefore very suitable for application by brushing, spraying or dipping.

The disadvantages are:

The cost. The solvents used are very much more expensive than water or oil. If, however, the solvents can be recovered the cost will be correspondingly reduced (see p. 116).
The solvents are generally flammable. Vapour in an enclosed space such as a roof may present a fire hazard unless precautions are taken to exclude the possibility of sparks from electric switches etc.

Emulsions

Emulsion preservatives, of which there are several on the market today, are designed for the *in situ* treatment of timber in buildings that have been attacked by woodworm or incipient decay but where the timber cannot be replaced without exorbitant cost. Relatively small quantities of preservative can be introduced into timber already built into a structure by brushing a fluid over its surface. Very much larger amounts can be introduced if a thick layer of a mayonnaise-like emulsion is laid over the surface. This breaks down slowly over several days and the preservative, thus released, diffuses into the underlying wood. Since the emulsion contains water diffusion of preservative can take place into wood that is too wet for an oil to enter.

Methods of Applying Wood Preservatives

A thin superficial treatment of the surface of wood with a preservative, however toxic it may be, can give only temporary protection. Apart from any loss of preservative from the surface which may take place as a result of evaporation or leaching, splits are likely to open up and allow spores to reach the untreated wood below and so set up decay in the interior. Superficial treatments, therefore, can only give satisfactory protection when the exposed surfaces can be regularly re-treated, as in weatherboarding or palings. But for timbers embedded in the ground, such as posts and poles, such treatment can do little to extend their life.

To get lasting protection of timber deep penetration of the preservatives must be achieved. When the timber contains both sapwood and heartwood the whole of the former should be permeated with the preservative fluid.

Superficial treatments include: application by brush; spraying and drenching; and dipping and steeping.

Impregnation treatments consist of: impregnation in pressure plants; open tank treatments; diffusion treatment of green timber; injection treatments; and sap replacement in freshly felled logs.

Superficial Treatments

Application by brush. This is a convenient method for applying a wood preservative to small areas such as garden sheds. Joinery and furniture infected with woodworm are generally treated by brushing an insecticide freely over the surface.

When applying preservative with a brush the wood should be flooded with the liquid to get the maximum absorption, and care must be taken to fill all cracks and splits. Tar oils should preferably be applied in hot weather, or should be heated if they are viscous at ambient temperatures. The ends of any timbers that have been cut to size after the treatment has been done should always be brushed over with preservative to protect the untreated wood that has been exposed.

Spraying and drenching. Spraying is a useful way of applying preservatives to large areas such as weatherboarding on houses and barns, and for applying insecticides to roof timbers infected with woodworm, as it enables operatives to reach timbers that are inaccesible to brush treatments. It is also convenient for applying fungicides to walls infected with dry-rot.

When spraying a preservative solution a nozzle should be used that is designed to deliver a coarse spray. Fine atomising sprays, as used in horticulture, are unsuitable as they tend to waste liquid and also to make the atmosphere in a closed space unpleasant for operatives.

Drenching with liquid in a spray tunnel is a mechanised method of spraying preservatives to finished woodwork but it does not achieve such good results as dipping.

Dipping and steeping. Total immersion in a bath of preservative is more effective than either brushing or spraying as it ensures that every part of the surface is completely wetted. If the period of immersion is quite short the absorption obtained is similar to that achieved by brushing, a ten-second dip giving about the same absorption as one brush coat. Soaking for ten minutes results in an uptake similar to that given by three brush coats.

Short periods of immersion may give worthwhile protection to thin material consisting mainly of sapwood, but for joinery such as window frames, a three-minute immersion in a solvent preservative should be considered the minimum while ten minutes would give more reliable protection.

Prolonged steeping for periods of a week or more can result in quite deep penetration into, for instance, seasoned posts. Round posts cut from softwood thinnings respond best to this treatment as they consist mainly of permeable sapwood. Obviously such treatments can only be used when small numbers of posts have to be treated as the tank has to be occupied by one charge for many days.

Impregnation Treatments

Impregnation in pressure plants. If timbers are merely immersed in liquid the penetration of preservative is mostly slow and irregular, so the idea was developed of applying an external pressure to force liquid into the pores of the wood.

The Full-cell Process, patented by Bethel in 1838, results in filling the cells of the treated zone with a liquid. This process involves loading the seasoned timber on to small bogies which are run into a steel cylinder that can be closed with a pressure door. Commercial cylinders vary in diameter from 3 to 6 ft, and they have been made up to 150 ft long. A vacuum is then drawn inside the cylinder to extract air from the cells of the wood. After an interval, depending on the size of the timbers being treated, the cylinder is filled with preservative. If a tar oil is being used this is generally heated to 75–90°C depending on the viscosity of the oil. A pressure of 1000–1200 kN/m² (= 150–175 lb/sq in) is then applied for a time depending on the kind of wood being treated and the dimensions of the pieces. After the required weight of preservative has been absorbed into the wood the liquid is drawn off and a short period of vacuum is applied to remove surplus liquid.

The Empty-cell Process is a pressure method for applying tar oils. This gives equally deep penetration but results in a lower final absorption of oil and minimises the risk of the oil 'bleeding' out subsequently. In this method the preliminary vacuum is omitted and the oil is applied to the wood either at atmospheric pressure (Lowry Process) or after a short pre-liminary air pressure has been applied (Rueping Process). In either case the air present in the wood becomes compressed and when the oil pressure is finally released this compressed air expands again and drives out some of the oil in the cell spaces, leaving a coating of oil on the cell walls.

A recent development of the empty cell process using organic solvent preservatives is the Vac-Vac Process. This is a process in which the degree of the initial vacuum and the degree and duration of the final vacuum are accurately controlled so that precise, previously determined, amounts of preservative can be put into the wood. Wood treated in this way is touch dry when it comes out of the cylinder and may be assembled without delay and primed shortly afterwards. Using Tn/BTO a loading of 0·1 per cent in the defined zones (about 0·48 kg/m³) is the minimum requirement. If a lique-fied petroleum gas is used as a solvent it penetrates the wood very easily under pressure. This solvent gas can be recovered by the vacuum subsequently applied, and then can be reused. In this so-called Drilon Process the treated wood contains no liquid and so does not require to be dried before it is painted.

The amount of preservative absorbed by the timber is

measured by weighing the charge before and after treatment. It is usual to express the degree of treatment in terms of weight of preservative absorbed per unit volume of wood— i.e. as kg/m³ (or lb/cu ft). For instance average retentions obtained in commercial practice vary between 96 and 190 kg/m³ (6–12 lb/cu ft), depending on the kind of wood and the dimensions of the pieces being treated. Higher absorptions are required for thin materials than for large sized timbers. It is extravagant and unnecessary to attempt to treat large sized timbers to the core. Provided there is a fully impregnated zone of adequate thickness on the outside of a pole infection cannot reach the untreated interior.

The minimum net retentions for creosote in different timbers for various purposes are laid down in BS 913 of 1973 from which the following examples are quoted.

Usage	Species	Minimum net retention kg/m³	Main. extended pressure period	Incising
Poles	European redwood	115	2	No
	Spruce	115	4	Desirable
Railway	European redwood	130	3	Optional
sleepers	Douglas fir	130	5	Essential
Marine piling	Elm	240	3	No
	Douglas fir	160	6	Essential
Fencing,	Permeable woods	110	2	No
domestic	Resistant woods	110	3	No
House timbers	Permeable woods	80	1	No
	Resistant woods	80	2	No

Premature failure of impregnated timber can usually be ascribed to insufficient seasoning of the wood before treatment. If the cells are already filled with sap or water the preservative fluid cannot enter them, and when these wet areas eventually dry out in service, the wood shrinks and cracks appear through which infection can gain entry and set up decay in these untreated areas.

Open-tank treatments. The cost of the complete plant for pressure impregnation of timber in cylinders is quite high and so for estate work in remote areas the Open-tank Process offers a useful and much cheaper alternative which gives adequate protection to permeable timbers. In this process the seasoned timber is submerged in a tank of preservative which is heated for a few hours and then allowed to cool while the timber is still under the liquid. During the heating period the air in the cells of the wood expands and much of it escapes as bubbles. When the timber cools again the air remaining in the cells contracts and the preservative is sucked into the pores.

A simple modification of the open-tank method is often used on farms and estates for treating the butt ends of posts. A stout steel drum of about 90 gallons capacity and measuring approximately 42 in high by 32 in wide, is placed on a rough hearth of bricks around a shallow hole in the ground, with a piece of 3 in wide piping fixed at the back to ensure a good draught. The posts to be treated, which must be well seasoned, are stood vertically in the drum which is then filled two thirds full with creosote. This is heated till it reaches about 200°F (93°C) at which temperature it is held for an hour or so before the fire is extinguished to allow cooling to take place. As oil is absorbed during cooling it may be necessary to add more preservative to keep up the level. At the same time the exposed parts of the posts above the oil should be brushed over freely with the preservative. Surprisingly good absorptions can often be obtained by this treatment if the timber treated is round, well-seasoned wood consisting mostly of sapwood.

Diffusion treatments. These depend on the ability of water-soluble salts to diffuse through wet wood. To get effective penetration by diffusion it is essential that the wood should remain in a moist condition for some time after the preservative has been applied in a concentrated form to the surface.

The most successful diffusion treatments for building timbers have been developed in New Zealand using concentrated solutions of sodium octaborate. The freshly-sawn green timber is dipped in highly concentrated hot solutions of the salt and then closely stacked in covered piles for a number of weeks depending on the thickness of the pieces treated.

Recommended concentrations, temperatures and diffusion storage times for various thicknesses are given in the Table below:

Thickness of timber		Solution conc. wt/vol boric acid	Min. operating temp	Min. diffusion period
mm	in	equivalent %	°C	weeks
25	1	25	46	4
38	1·5	30	52	6
50	2·0	35	55	8
77	3·0	45	57	10

Recent investigations by McQuire and Goudie, at the New Zealand Forest Research Institute, suggest that the time required for treating freshly-sawn Radiata pine with boron could be greatly reduced by a hot and cold bath followed by diffusion at elevated temperatures of about 50°C. Since only 7–9 days would be required, instead of 60–80 at ambient temperatures, the amount of timber held in diffusion at any one time would be reduced to about one tenth of that previously required for the same throughput. It is also claimed that better and more even treatment of the timber is achieved by this process.

After the diffusion storage the treated timber should be seasoned by air drying under cover, or by kilning. The treatment is deemed satisfactory if the average retention in the core of the pieces is not less than 0·3 per cent boric acid equivalent in sapwood, nor less than 0·05 per cent in heartwood.

By using two successive diffusion treatments, with different salts that react together to form an insoluble precipitate in the wood, it is possible to produce a preservative that is fixed and resistant to leaching. This so-called Double-diffusion Method can be used to achieve good penetration of preservative into such hard-to-treat species as Rocky Mountain Douglas fir. Suitable combinations of salts are copper sulphate and sodium arsenate, or copper sulphate followed by sodium chromate.

Injection treatments. An ingenious way of introducing preservative salts into poles is that known as the Cobra Process. A preservative paste, consisting of soluble salts, is injected through a stout, hollow, flat needle into the wood parallel to its grain to a depth of about 5 cm. The injections are usually made into a zone just above, or just below, ground level as this is where the risk of decay is greatest.

This method is most commonly used for the treatment of standing poles that have begun to show signs of incipient decay at the ground level. It is claimed that the average life of poles thus treated can be extended by 15–20 years. It can also be conveniently used for the initial treatment of poles in remote and inaccessible regions where no facilities for pressure impregnation exist.

Sap replacement. This method, by which freshly felled trees can be treated with water-borne preservatives has been called the Boucherie Process after the inventor who patented it in 1838. It was primarily intended for the treatment of poles.

The trees are laid, immediately after felling, in a horizontal position with rubber caps attached to their buttends. These are connected by hoses to a reservoir of preservative some 6 m or so above them. The hydrostatic pressure forces the preservative through the sapwood pushing the sap out of the top of the trunks. This process, employing solutions of copper sulphate, was at one time used extensively in France but was never very popular in this country. There has, however, been a revival of interest in it lately, using effective modern water-borne preservatives. In remote areas, where a supply of suitable, pole-sized timber is available and labour costs are not too high, this process can be economically advantageous.

Preparation of Timber for Preservative Treatment

Except for timber that is to be treated by the sap replacement process it is essential to remove all the bark from poles and any other timber that is to be treated in the round. The poles must then be piled openly so that air can circulate freely around them. The period required to dry down the sapwood to the required figure varies between 6 and 18 months according to size and to climatic conditions. A moisture content in the sapwood of about 25 per cent is suitable, and once all the free water has gone from the pores and cell spaces there is no advantage to be gained from further drying.

Sawn timber must likewise be seasoned down to a moisture content below the fibre saturation point before being impregnated.

It is important that any borings for bolts which are required for fixing—e.g. cross arms to poles or rails to sleepers—should be made before the wood is treated. Otherwise such borings may expose untreated wood in the interior and so become foci of decay.

Certain timbers, such as Douglas fir, that are resistant to the penetration of preservatives, can be satisfactorily treated only if their surfaces are incised. This process consists of making slits in staggered rows in the surface layers about $\frac{1}{2}$ in long and $\frac{1}{2}$ in deep (12 mm). Shallower more closely-spaced incisions can be used on smaller section sizes used in buildings. This permits a whole surface shell of wood to be adequately penetrated by preservative.

Water storage, or ponding, of logs before they are seasoned greatly increases the permeability of spruce poles. The action of bacteria in the water of the pond causes a significant increase in the permeability of the sapwood.

Specifications for Preservative Treatment

Standard specifications for the composition of certain well known preservatives, such as creosote, have been in existence for many years and should always be quoted where appropriate. But it must be emphasised that the method of applying a preservative is just as important as the nature of the preservative itself. Therefore any specification for treatment should state clearly how the preservative is to be applied and what retention of it by the wood is required. Merely to ask that some timber should be creosoted may result in just a single brushed on coat being applied, and this would be practically useless for timber that was to come into contact with the ground.

If pressure impregnation is required a permeable species of wood should be chosen—e.g. European redwood, not white-wood—and a minimum average retention of preservative per unit volume of wood should be laid down. If the timber is to be steeped then a minimum period of immersion should be demanded and if the preservative is to be applied by brushing

or spraying a minimum quantity of preservative per unit
surface area should be specified.

Fire Retardant Treatments

It is appropriate to discuss in this chapter treatments for
increasing the resistance of timber to fire.

Timbers vary greatly in the ease with which they can be
ignited and in the rate at which they will burn. Dense heavy
hardwoods are more resistant than light resinous softwoods.
Certain timbers, such as greenheart, gurjun, jarrah, iroko, oak
and teak, are classified as very resistant to fire and their use
for making fire-resistant doors is officially approved.

Impregnation of timber with certain salts greatly reduces its
flammability and prevents smouldering and flaming after the
source of heat is removed. But such treatments should not be
called 'fire proofing' as no treatment can render wood immune
to the effects of high temperatures—between 150 and 200°C—
which cause wood to char.

The methods used for impregnation with fire-resistant
chemicals are similar to those used for treating with wood
preservatives but much higher concentrations of salts are
required. Ammonium phosphate is now the preferred
chemical. A retention of 48 kg/m³ (3 lb cu ft) of the salt is
commonly specified for structural timbers. After impregnation
the treated timber must be redried, preferably in a kiln. If it is
to be used for joinery or cabin linings in a ship the moisture
content should be reduced to 8–10 per cent while for rafters
14–15 per cent would be suitable.

Although it is only by thorough impregnation with high
concentrations of the salt that timber can be rendered resistant
to prolonged heating, the rate of spread of flame can be
greatly reduced by surface coatings. There are two types of
these; those that act by preventing access of oxygen to the
wood, such as sodium silicate (waterglass) mixed with kaolin;
and those that swell up to provide in addition an insulating
layer. These so-called 'intumescent' paints can be based on a
mixture of urea formaldehyde with ammonium phosphate

plus a filler of an opaque finish. This type is unsuitable for outdoor use.

Fire retardants have their greatest value in preventing the outbreak of fire or for checking its rate of spread in the early stages when every minute counts. They are most useful for the protection of light timbers and insulating boards. There is no great advantage in treating large-sized structural timbers as their rate of burning is, in any case, quite slow. Wood is a poor conductor of heat, and though the surface of a beam may be burning, only a small proportion of the heat is transmitted through to the core which does not become hot enough to liberate flammable gases. It is for this reason that large timber beams retain their mechanical strength during a conflagration for much longer than unprotected metal ones which collapse when the metal reaches a critical temperature.

In certain situations, such as mines, it may be desirable to protect the timbers against fire and fungal decay together. It is a simple matter to include, in a mixture of fire retardant chemicals, salts that are toxic to fungi. Boron compounds are very suitable for this purpose as they themselves contribute both properties to the mixture.

BIBLIOGRAPHY

British Wood Preserving Association, London. Convention records and leaflets. 1947–1973.

Findlay, W. P. K. *The Preservation of Timber*. (London, A. & C. Black), 1962.

Hunt, G. M. and Garratt, G. A. *Wood Preservation* 2nd edition. (London, McGraw-Hill), 1967.

McQuire, A. J. and Goudie, K. A. *New Zealand Journal of Forestry Science* **2**, 165, 1972.

CHAPTER NINE

Identification of Timbers

Some woods can be recognised at a glance while others require detailed microscopic examination in order to establish their identity. It must always be remembered, when deciding from what tree a piece of wood has been cut, that the appearance varies considerably according to whether one is looking at the cross-section, the radial, or the tangential surface. For example, the characteristic silver grain of oak cannot be seen on the tangential surface. And, as already explained (p. 3), the quality of timber from trees of the same species varies greatly according to the conditions under which the trees have been grown. Timber from trees that have grown very slowly in the far north of Russia will look very different from that of the same species grown in the south of England.

Before proceeding to a detailed examination of the structure of the wood its general physical features should be noted, as for example:

Colour. Whether the sapwood and heartwood differ in this and whether variations in the form of streaks or mottling are present—as in walnut.

Texture. Whether coarse (e.g. oak); medium (e.g. walnut) or fine (e.g. box).

Feel. Some woods feel greasy or oily (e.g. teak and lignum vitae).

Density when dry. Whether light weight—sp. gr. below 0·5; medium—0·5–1·0; or heavy, sp. gr. more than 1·0, when it sinks in water even when air dry.

Odour of seasoned wood. A few timbers, such as Western red cedar, have a highly characteristic smell. This is best noted when a fresh surface is exposed.

Hardness. Testing with the finger nail can indicate whether the wood is hard, medium or soft.

Colour of Ash—made by burning small match-sized pieces. This is a useful test for distinguishing jarrah which produces black ash from karri which has white ash.

To identify wood from its structure it is necessary first to examine an absolutely cleanly-cut cross-section of the end grain under a 10× lens. Next the radial surface can be examined by splitting the wood along the plane of the rays. A very large number of hardwoods can thus be identified without recourse to the compound microscope. But there are many coniferous softwoods that require microscopic examination to ensure precise identification. It is, of course, helpful, to know the region from which the timber in question has come as this limits the possibilities. But it must be borne in mind that some species, such as Radiata pine and *Eucalyptus* spp. for example, have been widely planted in countries remote from the places where they grew naturally.

Identification of Softwoods

The principal features to be looked for when attempting to identify a softwood are:

The relative density and proportions of the early and late wood zones. In some woods, such as Douglas fir, the latewood zone is very conspicuous with a fairly abrupt change from earlywood to latewood. But in others, such as Sitka spruce, the latewood zone is less conspicuous and the wood has a more uniform texture.

Resin ducts. These appear under a lens as occasional small spots or holes on the end surface of pines, spruces, Douglas fir and larch. On a freshly-planed, clean, radial section of these species they appear as streaks or lines.

TABLE 3.   Identification of softwoods

	Annual rings	Resin ducts	Other features
Larch	Strongly marked by contrasting spring-wood and summer-wood.	Present but barely visible to the naked eye.	Heartwood reddish-brown; resinous. Sapwood narrower than in pine.
Scots pine and Baltic redwood	Ditto	Generally visible on clean-cut end-surface or freshly planed longitudinal surface.	Heartwood pale reddish-brown or yellowish-brown; resinous. (Corsican pine is similar but has a wider sapwood).
Douglas fir ..	Ditto	Present but barely visible to the naked eye.	Heartwood reddish-yellow to brown; moderately resinous.
Western hemlock	Clearly marked by moderately dense summerwood.	Normal resin ducts absent. 'Black streak' is due to bands of abnormal resin ducts.	Pale greyish brown, somewhat darker than Sitka spruce; non-resinous.
Spruce	Not so strongly marked as above. Summerwood less dense.	Fairly distinct but comparatively sparse.	Light yellowish-brown (almost white) throughout; slightly resinous. Sitka spruce has a light pinkish-brown heartwood.
Noble fir ..	Similar to spruce.	Absent.	Similar to spruce, but not so lustrous; sometimes marked with brown resin streaks.
Western red cedar	Clearly marked by moderately dense but narrow bands of summerwood.	Absent.	Distinctive reddish-brown colour often streaked; non-resinous. Distinctive odour. Lighter in weight than other commercial softwoods.
Parana pine ..	Comparatively indistinct. Little contrast between springwood and summerwood.	Absent.	Pale brown often with peculiar red streaks. Non-resinous. Distinguished also by its fine even texture.

Reproduced from FPRL leaflet No. 34. C.C.R.

Identification of Hardwoods

The presence of annual growth rings is a feature of most temperate-zone hardwoods, but many tropical hardwoods do not possess them. The principal features to be looked for when identifying hardwoods are:

The distribution and grouping of the pores; e.g. whether the wood is ring porous or diffuse porous.
The size and distribution of the rays.
The distribution of any soft tissue (parenchyma).

How these three features can be used to identify some of the common British hardwoods is illustrated in the table below. When choice has to be made between only a relatively small number of species a dichotomous key such as this may be helpful. When using this type of key one proceeds from the first question to the next, following the numbers as they apply, until the answer is arrived at.

Key to Common British Hardwoods

1   Pores of earlywood in a conspicuous ring or
     band?   Yes, (i.e. ring porous) 2
                           No, (i.e. diffuse porous) 5
2   Pores of latewood in radial lines or group   3
      Pores of latewood in scattered or tangential
         lines   4
3   Rays of two distinct sizes, larger ones very
         broad   Oak
      Rays all very narrow   Sweet chestnut
4   Pores of latewood single or in scattered
         groups sometimes joined by soft tissue   Ash
      Pores of latewood in distinct wavy tangential
         bands   Elm
5   Pores individually visible to the naked eye?
                                      Yes, Walnut
                                       No, 6
6   Rays of two distinct sizes, larger broad and
         conspicuous?   Yes, 7
                                       No, 8
7   Pores in radial rows, broad rays faintly
         visible as long lines in tangential face   Alder

	Pores not in radial rows, broad rays clearly visible on tangential face	Beech
8	Rays individually distinct to the naked eye?	Yes, 9
		No, 11
9	Rays more than twice as wide as pores?	Yes, Plane
		No, 10
10	Wood white or yellowish; annual rings marked by a white line	Sycamore
	Wood light brown or reddish; annual rings shown by abundant pores in the early-wood	Cherry
11	Wood hard and fairly heavy	12
	Wood soft and moderately light	13
12	Wood pale coloured, lustrous	Birch
	Wood reddish, not lustrous	Apple; Pear
13	Rays fairly distinct with lens?	Yes, 14
		No, Willow or Poplar
14	Rays irregularly spaced, separated by width of 2 or more pores	Lime
	Rays closely spaced	Horse chestnut

While a dichotomous key can usefully be constructed to include the principal timbers of a region, or to assist in recognition of the various related timbers in a botanical family, it is impossible to prepare a key of this sort which will include all the commercial timbers of the world. Another disadvantage of this type of key is that, once prepared, additional species cannot be included without rewriting large sections of it.

The use of marginally-perforated cards to record the dianostic features of timbers, which can then be selected in any sequence to permit rapid diagnosis, was first developed in 1936 by S. M. Clarke at the Forest Products Research Laboratory at Princes Risborough, and has since been adopted in many other timber research laboratories.

In this system the diagnostic features for each timber are recorded by clipping out the appropriate marginal holes corresponding to the features present in the wood (see Fig. 28).

Some species show little variation in the anatomy of their

SOFT TISSUE (PARENCHYMA)

K	ABSENT OR INDISTINCT	23
L	DISTINCT TO NAKED EYE	24
M	TERMINAL PROMINENT	25
N	APP. TERMINAL ONLY	26
O	PREDOM. INDEP. OF PORES	27
P	DIFFUSE	28
Q	PREDOM. ASSOCD. WITH PORES	29

SURROUNDING PORES	30
WING-LIKE OR CONFLUENT	31
BANDED	32
BROAD. CONSPIC. BANDS	33
FINE LINES	34
RETICULATE	35
LADDER-LIKE (SCALARIFORM)	36
	37

OTHER FEATURES

INCLUDED PHLOEM	38
VERTICAL CANALS	39
Oil or Mucilage Cells	40
Fibres Radial	41
STORIED (NOT RAYS)	42
	43
	44

PORES (VESSELS)

J		
22		> ½ WIDTH OF VESSELS 45
21		WIDER THAN VESSELS 46
20	SIZE 6	47
19	SIZE 5	SIZE 3 OR LESS 48
18	SIZE 4	SIZE 4 49
17	SIZE 3	SIZE 5 50
16	SIZE 2 OR LESS	SIZE 6 51
15	" " " < 250 "	SIZE 7 52
14	" " " 125-250	53
13	" " " 65-125	Heterogeneous 54
12	" " " 30-65	" : Several Marg. Rows 55
11	" " " 12-30	> 2 MM. IN HEIGHT 56
10	< 12 Per 10 sq. mm.	Canals or Latex Tubes 57
9	DEPOSITS OR GUM	STORIED 58
8	TYLOSES ABUNDANT	59
7	Simple Perforations	60
6	Multiple Perforations	< 25 Per 5 mm. 61
5	Pore Clusters	25-50 " " 62
4	TANGENTIAL ARRANGEMENT	50-80 " " 63
3	RADIAL OR OBLIQUE	> 80 " " 64
2	RADIAL MULTIPLES	3 OR MORE PER VESSEL 65
1	EXCLUSIVELY SOLITARY	AGGREGATE RAYS 66

RAYS

NAME:—

FAMILY:—

F.P.R.L. Lens Key to Hardwoods.

	GRO. RINGS	GEOGRAPHICAL REGIONS											
H													
G	RING POROUS												
F	COUNTABLE												
E		TEMP. SOUTH AMERICA.											
D		TROP. AMERICA. W. INDIES											
C		NORTH AMERICA											
B		SOUTH AFRICA.											
A		TROP. AFRICA. ETC.											

PHYSICAL PROPS.	
HEAVY:	74
MEDIUM: " : 0.5-1.0	74
MEDIUM: " : > 1.0	75
LIGHT: SPEC. GRAV. < 0.5	73
SPLINTER BURNS TO ASH	72
REDDISH	71
YELLOWISH	70
WHITE	69
DISTINCTIVE COLOUR	68
DISTINCTIVE ODOUR	67

GEOGRAPHICAL REGIONS: AUSTRALIA, NEW ZEALAND 79, MALAY, ETC. 80, INDIA. ETC. 78, EUROPE. ETC. 77, NORTH AMERICA 83, SOUTH AFRICA. 82, TROP. AFRICA. ETC. 81, TEMP. SOUTH AMERICA. 85, TROP. AMERICA. W. INDIES 84, COUNTABLE 86, RING POROUS 87, 88

28. Card for recording features of hardwoods under lens. C.C.R.

wood and a single card can adequately describe them but for others two or more cards may be required to cover the variations observed. To identify an unknown timber the pack of cards on which the descriptions are recorded is sorted by inserting a needle of appropriate size through the hole corresponding to the first feature observed. The whole pack is then shaken and all the cards in which the hole corresponding to this feature has been clipped, fall out. Another and then another feature is selected and each time fewer cards drop out until finally one card is left which, if the selection has been done correctly, gives the identity of the timber under consideration.

An important advantage of this type of key is that any obvious or unusual feature can be used to begin the selection, and features can be used in any sequence. Selection can also be done in a negative sense. Timbers in which a character is absent are those which remain on the needle when the pack is shaken.

The diagnostic features that permit identification of some 400 hardwoods by this method are described and listed in the Forest Products Research Bulletins referred to in the Bibliography. From these bulletins students can themselves prepare sets of cards that will enable them to identify timbers once they have become familiar with the meaning of the terms and are able to recognise the features. Generally keys are based either on features that can be observed under a 10 × lens or on those that require microscopic examination of sections cut in the three main planes—transverse, radial and tangential; but there is no reason why a key should not combine both types.

As with the recognition of any biological specimens, be they plants, insects, birds or timbers, practical experience is required. The best way to learn to recognise timber is to examine authentic specimens of the woods themselves, comparing them with their written descriptions and using the available keys. The student must always remember to use not only his eyes but his nose and fingers, and at the same time to bear in mind any information he has about the origin of the specimen.

It is often very important for practical and commercial reasons for a buyer to be able to check whether the timber supplied is in fact the species for which he contracted. For example, if softwood fence posts are to be creosoted it is most desirable that the wood supplied should be European redwood and not whitewood (spruce), the latter being resistant to impregnation; and again if oak staves for barrels are required these must be made from an impervious white oak and not from a porous red oak. Therefore while precise identification of all the timbers in the international trade is something that only a trained wood anatomist can hope to achieve, it behoves every user of timber to learn to recognise the woods with which he is personally dealing.

Commerce apart, many people have found the recognition of timbers an interesting and challenging study, and indeed for some it has the fascination that philately, and other such hobbies, have for others.

BIBLIOGRAPHY

Identification of Hardwoods. A Lens Key. Forest Products Research Bulletin No. 25 (London, H.M.S.O.), 1952.

Brazier, J. D. and Franklyn, G. L. *Identification of Hardwoods. A Microscopic Key.* Forest Products Research Bulletin No. 46 (London, H.M.S.O.), 1961.

CHAPTER TEN

Principal Timbers of Commerce and their Sources

The firms that deal in commercial timbers have traditionally been divided into those selling softwoods or hardwoods, but today many of the large firms which have been formed by amalgamations deal in both. The trade is also divided into the merchants who deal primarily in imported timbers and those who buy home grown logs and convert them in their own mills.

The pattern of the world trade in timbers is continually changing and some timbers that were formerly available in large quantities are today unobtainable, the forests that supplied the logs having been felled and the land replanted with species that mature more quickly. This has happened in New Zealand where radiata pine has largely replaced the native hardwoods. In some countries replanting of forests has been sadly neglected and woodlands that formerly carried useful trees now bear only weed species which have little value as a source of timber, or else the land has been converted to agricultural uses. There are however certain timber exporting countries such as Sweden and Finland that now wisely restrict their fellings so that the volume of timber felled in any one year is approximately equal to the increase in volume by the growth of the standing forests. Another satisfactory method of forest management used in some places is to fell each year only that proportion that can be recovered by replanting. Thus if it takes sixty years for a tree to reach merchantable size then only one sixtieth of the forest acreage should be felled in any one year. In those countries in which felling is not allowed to exceed the annual gross increment a continuity of supply is thus ensured.

In spite of the substitution of other materials for a number of purposes the world demand for timber continues to increase as the population and its standard of living rises. It is likely therefore that the price of timber will increase more than that of other raw materials, such as cotton for instance, larger crops of which can rapidly be grown in response to greater demand.

Naming of Timber

When only a few native timbers, such as oak, beech and pine, were used their names were well known and could not be misunderstood, but as the international trade increased so new names were given to the familiar timbers if they had been imported from abroad. Scots pine, or Scots fir as it was often called, when imported from Scandinavia became known as European redwood, or red deal. Similar but not identical timbers such as oak from America would be called American white oak, or American red oak, while a timber which bore only a superficial resemblance to true oak, such as so-called Tasmanian oak, might be mistakenly expected to possess similar strength and durability to the English oak.

This multiplicity of names often led to misunderstandings and to the use of unsuitable timbers because of uncertainty as to their true identity. A list of standard English names has therefore been drawn up by the British Standards Institution and published as BS 881, 589. All the names used in this book conform to the BS list.

The botanical Latin name of a plant or tree is, according to the rules of nomenclature, the first validly published name given to it either by Linneaus in 1753 or subsequently. These names normally consist of two words, the first signifying the genus to which the plant belongs and the second the species;

e.g. *Pinus sylvestris* (the Scots pine). For some newly introduced timbers no English name exists and it is necessary to use a botanical name to ensure precision. There may exist a dozen or more local native names for the same timber if it is exported from a number of ports. For example thirteen alternative names are given for the wood *Lovoa klaineana*, the so-called African walnut—which anyhow is not a true walnut at all!

Unfortunately botanists still sometimes change the Latin names, either because there has been a confusion as to the identity of the tree when the name was first given, or because an earlier name has been discovered in a hitherto unnoticed publication. The Latin name of the Douglas fir, which at present has been settled as *Pseudotsuga menziesii*, has been changed at least four times in the author's lifetime.

Softwoods

The softwoods described below are those most commonly used at the present time in the United Kingdom. The most important sources of supply mentioned here are those that apply at the time of writing; but it must be emphasised that these may change as the prices rise in various countries or political and economic factors intervene to affect the supply from any particular region.

Summarised data about the physical properties, resistance to decay, permeability to impregnation with preservatives, and working properties of the principal softwoods used in Great Britain are given in Table 4.

Douglas fir. This tree is sometimes called 'Columbian, or Oregon, pine'. It is not a true fir since it belongs to the species *Pseudotsuga*. The variety imported from Canada is one of the few conifers from which large baulks in long lengths can still be obtained. The timber from home grown trees is, however, of more rapid growth and of coarser texture than that of the imported wood.

The heartwood of Douglas fir is a light reddish-brown with strongly marked annual rings due to the contrast between the earlywood and latewood zones. This gives rise to a prominent growth figure in the rotary cut veneers.

This is one of the most useful timbers for heavy constructional work, such as shores for buildings and roof trusses. It makes excellent vats and tanks. If used for flooring it is important that only rift-sawn (edge grain) material should be chosen because the annual rings tend to flake away and splinter if plain sawn boards or blocks are used.

The timber from this tree can be succesfully impregnated with creosote only if it is incised before treatment (see p. 121).

Silver fir. There are several species of true fir (*Abies* spp.) grown in this country. The only one that is imported in any quantity is the European silver fir which is sometimes called whitewood and is often mixed with European spruce in shipments of 'whitewood' from central Europe.

This timber closely resembles that of European spruce being almost white, or pale yellowish-brown, but it is somewhat less lustrous. It seasons rapidly and well with little degrade. It is used for joinery and general carpentry work and in Scotland it is the preferred timber for domestic floors. Large quantities are also used for packing cases.

Larch. Most larch used in the United Kingdom comes from home-grown trees either of the European or Japanese species. It is the hardest and toughest of the ordinary home-grown softwoods. The resinous heartwood is pale reddish-brown or brick red and is sharply differentiated from the narrow band of sapwood which is pale-coloured.

Larch is the most durable of the commonly grown softwoods and if straight grained makes excellent planking for boats. Many Scottish fishing boats have hull planking made of larch. Thinnings are often used for fencing and poles and it is a favourite timber for gates, sheds and other such purposes where durability is a very important factor. It seasons rather more slowly than Scots pine.

Parana pine. This is not a true pine but is a relative of the monkey-puzzle tree. Its timber, which is imported from Brazil, has a brown heartwood in which narrow bright-red streaks often occur along the grain; sometimes these may be found at the edge of the paler sapwood. The timber shipped to this country is generally free from knots and is available in long wide boards which makes it a very suitable wood for shop fittings and joinery. It is not however durable and should

TABLE 4.

Standard name	Latin name	Av. density kg/m³	lb/cu ft	Strength relative to redwood	Durability of heartwood	Permeability to preservatives	Working properties
Douglas fir	Pseudotsuga menziesii	528	33	Stiffer and harder 30% stronger in bending	Moderate	Resistant	Works readily but rather harder than redwood
Silver fir (Whitewood)	Abies alba	480	30	Similar	Non-durable	Moderately resistant	Works easily, but requires very sharp tools
Larch (European)	Larix decidua	592	37	50% harder and rather tougher	Moderate	Resistant	Can be sawn and machined fairly readily; finishes cleanly
Parana pine	Araucaria angustifolia	544	34	Similar but less tough	Non-durable	Moderately resistant	Works easily; finishes well
Pines (true) Caribbean pitch	Pinus spp. P. caribaea	704	44	Stronger, harder and denser	Moderate	Moderately resistant	High resin content makes some operations rather difficult
Corsican	P. nigra	512	32	Similar	Non-durable	Moderately resistant	Works well if not too knotty
Radiata	P. radiata	480	30	Similar	Non-durable	Permeable	Works easily
Scots (Baltic redwood)	P. sylvestris	512	32	Similar	Non-durable	Moderately resistant	Depends partly on country of origin

Spruce European	*Picea abies*	464	29	Similar	Non-durable	Resistant	Works easily; finishes cleanly
Sitka Imported	*P. sitchensis*	432	27	Similar—high for its density	Non-durable	Resistant	Works easily. If home grown requires very sharp tools
Home grown		400	25	25% stiffer	Non-durable	Resistant	
Western hemlock	*Tsuga heterophylla*	480	30	Similar	Non-durable	Resistant	Works readily
Western red cedar	*Thuja plicata*	368	23	20–30% weaker, also softer	Durable	Resistant	Works easily but tools must be very sharp
Yew	*Taxus baccata*	672	42	Harder, tougher, more resilient	Durable	Resistant	Works to a good finish. Turns excellently

not be used out of doors or for any purpose where toughness is particularly required, as, for instance, in ladder sides or scaffold planks.

Pines

Only a few of the 96 or so species of pine are much used as sources of timber in the United Kingdom but those few yield some of the most important timbers in the northern hemisphere.

Pines can be classified into three groups according to whether the needles grow together in pairs, in threes, or in fives. The two-and three-needle pines yield timbers in which the earlywood and latewood zones are fairly clearly distinguished with well marked annual rings, as in the Scots and Corsican pines which have two needles and the pitch pines (such as the longleaf pine from the U.S.A. and the slash pine from Central America) which have three. In the five needle pines, on the other hand, the annual rings are not at all clearly marked and the wood, therefore has a much more even texture. Yellow (Weymouth) pine (*P.strobus*) and Western white pine (*P.monticola*) are examples of this type.

Corsican pine (Pinus nigra). This tree has been planted extensively in England as it grows more rapidly than the native Scots pine. The timber is very similar in its properties to the latter and it is used for similar purposes, but the sapwood is much wider and the texture somewhat coarser. This wide sapwood is useful if deep penetration of wood preservative during impregnation is required.

Maritime pine (Pinus pinaster). This is a native of the coastal regions of Southern Europe and is the principal tree in Portugal from where considerable quantities of sawn timber, in the form of pallet boards, box woods and fencing, are imported to this country. The wood is generally similar to that of Scots pine but is more resinous and generally knottier and rather coarser. Some samples that contain a great deal of resin may be appreciably heavier than Scots pine.

Radiata pine (Pinus radiata). This tree, which is not an important source of timber in the U.S.A. where it grows wild, has been planted over large areas in New Zealand and South

Africa. In those countries it grows very rapidly producing trees that contain little heartwood by the time they are felled. They are used as building timbers, for which purpose they usually receive treatment with a water-borne preservative. Large quantities are also used in the pulp mills of New Zealand to produce Kraft and newsprint papers.

Scots pine, or European redwood (*Pinus sylvestris*). This is one of the few really native conifers in Great Britain and at one time forests of it covered great areas of Scotland. During this century it has been planted extensively in East Anglia. Over much of Northern Europe it is the principal source of softwood and large quantities of Baltic redwood (as it is often called) are shipped annually to the U.K. from the Baltic and from Russia. Its timber has come to be regarded as the standard by which other softwoods are judged, though its quality and properties vary considerably according to the area in which the trees have been grown. The wood from northern Sweden and Russia has a fine texture on account of its very slow rate of growth and it is well suited for the manufacture of joinery. That grown in southern Sweden and the U.K. is coarser but is useful for carcassing and other building purposes. Its working properties and uses depend to a large extent on the number and size of knots that it contains (see Chap. 15 on grading). It is very suitable for transmission and telegraph poles as the broad band of sapwood is readily permeable and so can be effectively impregnated with creosote.

As the present day supplies of European redwood are coming from relatively small trees most of the boards contain a fairly high proportion of sapwood which is both porous and susceptible to insect and fungal attack. It is therefore advisable that all joinery and roofing timbers made from this timber should be pretreated with wood preservative before being incorporated in a building.

Spruce (*Picea spp*)

There are a large number of *Picea* species widely distributed all over the world. Two of these are of considerable economic importance in the U.K., the European whitewood—as it is commonly called—and the Sitka spruce from Canada.

All the spruce yield pale coloured softwoods in which the latewood zones are less pronounced than in the pines so that the texture of the timber is fairly uniform.

European spruce or whitewood (Picea abies). This tree is widely distributed in Europe, extending from Russia to the Pyrenees. In the young stage it is sold as the Christmas tree. Commercial supplies come from northern Europe and from western Great Britain. The wood is a very pale yellowish brown, or almost white, with no visible heartwood. Though it contains resin ducts it is never very resinous. The wood is used for similar purposes to European redwood and in Scotland is particularly in demand for domestic flooring. Smaller trees are used for scaffold poles, ladders, masts and pit props. In Scandinavia it is extensively used for making paper pulp. Being resistant to impregnation it is not as suitable for poles as redwood.

Sitka spruce (Picea sitchensis). The virgin stands of this tree in western Canada yield timber of fine quality which is strong for its weight, being 25 per cent stiffer than European redwood. At one time it was used extensively for construction of aircraft frames and gliders and it is still used for making oars and racing boats, and for sound boards in pianos.

This tree grows rapidly in the wetter regions of the U.K. and has been planted extensively by the Forestry Commission. The timber from these rapidly grown young trees is not so strong as the imported wood. The lower grades are suitable only for packing cases and for pulping but the better grades of rather denser material can be used for rafters and joists in buildings. The very fast grown wood has a tendency to collapse and warp during drying.

Western hemlock (Tsuga heterophylla). This tree, which grows in the west coastal regions of America, yields a straight-grained wood of even texture, pale brown in colour, with well marked annual rings. When freshly felled it has an unusually high moisture content and it does not season as quickly as most other softwoods, so the centre of thick planks may retain a high moisture content for a long time after they have been sawn. The timber, which can usually be obtained in large sizes, is used for all kinds of construction work but it is not durable and, being resistant to impregnation, should not be used in exposed situations.

Western red cedar (*Thuja plicata*). Commercial supplies of Western red cedar are imported from British Columbia though there are a few plantations in this country from which limited quantities are sometimes available.

The mature trees yield a sweet-smelling, non-resinous, light weight and rather soft timber. The heartwood varies considerably in colour from salmon pink to darkish chocolate brown. The narrow band of sapwood, which is not usually present in imported sawn timber, is whitish. The wood is usually straight grained and it seasons well with little degrade or distortion. It works easily with well sharpened tools but may be subject to 'chip bruising'. It contains an extractive called thujaplicin which is very toxic to wood-rotting fungi and this confers an unusually high durability on the timber. It is used extensively for making greenhouses and garden frames. It is very popular also for cladding timber houses and in a clean atmosphere it weathers to a pleasing silver-grey colour. Wooden roofing shingles are now usually made from this tree but these should receive preservative treatment if a life service of over 15 years is required of them.

Western red cedar has acidic properties and corrodes ferrous metals causing black stains so galvanised nails should always be used to fix boards and shingles made from its wood.

Yew (*Taxus baccata*). Though technically a softwood because it is a conifer, the yew is one of the heaviest and hardest of the softwoods and has many of the properties of a hardwood. Its hardness and compressive strength are similar to those of oak. The tree grows slowly and on account of its branching habit seldom forms a good bole. The narrow ring of sapwood is almost white while the heartwood varies from orange to dark purplish brown with well marked annual rings which give rise to a very pleasing figure.

This wood bends well and turns excellently and it takes a high polish. It is often used to make high-class chairs and ornamental bowls. As it is very durable in contact with soil the rougher pieces make good fence- and gate-posts.

Hardwoods

There is a greater diversity among hardwoods than softwoods and there are few useful generalisations that can be made about their properties. Though many are highly durable it is a mistake to assume that they all are, and the term 'hardwood' should never be used in a specification when what is meant is a timber having properties similar to those of oak.

Broadly speaking hardwoods can be grouped into those, such as oak, that are used for constructional purposes and furniture, and those used for specialised or purely ornamental purposes, such as lignum vitae, boxwood and ebony. There are of course many woods that were once used for general purposes but now, owing to their scarcity, are generally sliced into thin veneers for a finish to less valuable woods or chipboard.

General List

Abura (Mitragyna ciliata). This is a light-weight hardwood from West Africa, pinkish-brown in colour without any pronounced figure. The wide sapwood is not clearly differentiated from the heartwood. Its strength properties are similar to those of European redwood but it is somewhat harder. Its average density is about 560 kg/m³ (35 lb/cu ft). It is obtainable in fairly large sizes and is very suitable for utility furniture, interior cabinet work and light construction. It can readily be stained and polished; it works fairly easily, and takes thin gauge nails satisfactorily. It is not durable and should not therefore be used out of doors or in positions where it will be exposed to persistent damp.

African Walnut (Lovoa klaineana). This timber which comes from large trees in West Africa has a yellowish-brown colour and is sometimes marked with dark streaks similar to those in the true walnut—hence the name. Also the interlocked grain gives a well marked ribbon or stripe figure when logs are cut on the quarter. Its strength properties are similar to those of American walnut but it is somewhat weaker in bending. Its average density is 544 kg/m³ (34 lb/cu ft). The timber works easily, finishes cleanly and polishes well. It is used mainly for panelling, shop fittings and decorative joinery.

Afrormosia (Afrormosia elata). This wood is imported from Ghana and the Ivory Coast. Being extremely durable it has found a ready market in this country as a substitute for teak which it somewhat resembles. The dark yellowish-brown heartwood is fairly straight grained. Its average density when seasoned is 688 kg/m³ (43 lb/cu ft). Its strength properties are similar to those of European beech. It can be finished cleanly and polishes well but it tends to split if nailed and it requires care in machining operations. It is eminently suitable for high-class joinery and it can be used for ships' decking; but it will stain in contact with ferrous metals.

Agba (Gossweilerodendron balsamiferum). The large trees from which agba is obtained grow mainly in Nigeria. It is a light-weight hardwood of a uniform pale brown colour. It is generally straight grained and it has a finer texture than most African mahoganies to which its strength properties are similar. Its average density is 510 kg/m³ (32 lb/cu ft). It is likely to develop brittleheart (see p. 49) and therefore must be carefully excluded if strength of wood is an important consideration. It works easily, though occasionally gum collects on the teeth of the saws during conversion of logs. It is a useful timber for joinery components, both interior and exterior, because it combines easy working with natural durability. It is very suitable also for making light coloured furniture and polishes well if the grain is properly filled.

Ash (Fraximus excelsior). This is one of the most important of British hardwoods on account of its remarkable toughness and of the ease with which it can be bent. The quality of the timber depends to a considerable extent on the conditions under which it has been grown. The trees require a good deep moist soil if they are to grow really well. Being a ring-porous wood, slow-grown trees form wood of low density and the best quality comes from trees that have grown rapidly forming wide annual rings with plenty of dense latewood. The strongest ash usually has between 6 and 10 rings to the inch. The density thus varies considerably with an average of 688 kg/m³ (43 lb/cu ft). The wood is generally light coloured and sometimes is almost white. No heartwood is normally formed though some trees develop a dark brown central zone which is not, as might be thought, due to incipient decay. The timber is typically straight grained and works fairly easily. The general strength properties are similar to those of oak but it is much more resistant to splitting. Its toughness (shock resistance) is exceptionally high and on account of this it is the best timber for sports goods, such as hockey sticks for example, and for gymnasium equipment such as parallel bars. It is also very useful for the handles of such tools as hammers and axes, and it has been used extensively for agricultural machinery that will be subjected to hard wear and repeated stresses. Formerly large amounts were used for coach building and in the construction of bus and lorry bodies.

Beech (*Fagus sylvatica*). Beech, both home grown and imported, is used in larger quantities in the United Kingdom than any other hardwood. In the British Isles it is most prevalent in eastern Scotland and southern England, growing best on calcareous soils such as are found in the Chilterns and Gloucestershire. It grows also throughout central Europe and has been imported from Denmark, France, Yugoslavia, Roumania and other countries.

The wood of the beech is straight grained with a fine even texture, pale coloured at first but darkening to a light reddish-brown. The steaming treatment often given to it in central Europe renders it an even darker colour. The wood from northern countries tends to be harder and heavier than that from southern Europe. Home grown beech has an average density of 720 kg/m³ (45 lb/cu ft). It requires care during seasoning as freshly sawn boards are liable to warp and split if dried too quickly. It shrinks considerably on drying and shows large movement as a result of moisture changes. It is one of the strongest of home grown timbers and after seasoning has somewhat superior strength properties to oak. It bends exceptionally well.

Beech is not durable and logs must be extracted from the woods soon after felling otherwise they become affected by 'dote' (incipient decay). It is permeable and large quantities have been used in Europe for sleepers after impregnation with creosote. The wood is immune to powder-post beetles but susceptible to furniture beetle. It has been, and still is, used in large quantities for making furniture. The furniture industry in High Wycombe developed because ample supplies of beech were available from the nearby Chiltern beech woods. Good plywood can be made from it, owing to its close texture, it is an ideal wood for turnery and is used for tool handles, shoe heels, toys and brush ware. In wartime it was used for rifle butts.

Birch (*Betula pubescens*, or *B.verrucosa*). There are not many birch trees with good straight stems in Great Britain but in Scandinavia many trees have clean cylindrical boles which are used for the production of veneers for plywood. The whitish wood is fairly straight grained and fine textured without obvious annual rings or heartwood. It is hard and tough, comparable in strength properties to oak but considerably tougher. The average density is 660 kg/m³ (41 lb/cu ft). It is not at all durable and is liable to deteriorate if left for long in the woods after felling, but it is permeable to wood preservatives and, if properly impregnated with creosote, will last well as fence posts.

Home-grown birch is generally felled in the pole stage and used for making brush backs and bobbins etc. The imported timber is used for furniture making.

Cherry (*Prunus avium*). Wild cherry trees are scattered throughout woodlands on the better soils of this country. The wood has a fairly fine texture with whitish sapwood and a pale, pinkish brown heartwood which darkens after polishing to a pleasing colour resembling that of mahogany. Its strength properties are similar to those of oak but the wood is not durable enough for use in exposed situations. It turns well and takes polish excellently and so is very suitable for furniture and is also used for domestic ware and toys.

Sweet chestnut (*Castanea sativa*). Commercially sweet chestnut is normally grown as coppiced trees to provide pole sized material for cleaving into split pales for fencing. Parkland trees can reach a great size and yield timber that is very similar in appearance to oak but which lacks the large rays which give to the latter its silver grain figure. It is also much less strong and is difficult to season without collapse. Large trees often contain spiral grain and ring shakes which make conversion difficult. The timber is durable in the ground and hence is used much for fencing and gates. It is also used for furniture and coffin boards.

Elm (*Ulmus spp*). The elm tree often grows to a large size in hedgerows but has rarely been grown in plantations. There are several species in Great Britain, the timbers differing somewhat in their properties. There is a clear distinction between the pale sapwood and the dark, somewhat reddish-brown heartwood. The annual rings are conspicuous and their irregularity combined with the cross-grained nature of the wood give rise to an attractive figure. On account of the irregular grain the timber tends to distort on drying but this tendency can be reduced by weighting down the seasoning piles with blocks of concrete. The strength properties are about 30 per cent inferior to those of oak, but owing to its cross-grained nature it is difficult to split and therefore takes nails well. Wavy-edged elm boards are often used for cladding to give a rustic appearance to country houses and sheds. It is used for the manufacture of furniture, particularly by one firm who have made a speciality of elm furniture and use little else; and traditionally it has been the timber used for the seats of Windsor chairs. Large quantities are also used for coffin boards. Owing to its reputation for durability in water it has been used for the underwater parts of boats such as the deadwood and hogs in yachts and the bottom planking of canal boats. In the Middle Ages hollow elm pipes were used as water conduits. Nowadays it is much in demand for the manufacture of pallets.

Owing to the large numbers of elms being felled as a result of the depredations of Dutch elm disease considerable quantities of this wood are now available in England and additional uses for it should be sought.

Rock elm (*Ulmus thomasi*). The timber of this Canadian tree is much tougher than English elm and as it bends well it has been used a great deal for frames in boats. It is not as resistant to decay as oak.

Greenheart (*Ocotea rodioei*). Supplies of greenheart come from Guyana. It is a heavy timber of outstanding strength and durability and is obtainable in large, long baulks. The heartwood is usually a dark olive green and has a fine even texture free from knots and defects. The average density of the seasoned wood is 1030 kg/m³ (64 lb/cu ft). As it is resistant to rot and to marine borers it is particularly suitable for marine piling and harbour works and also for keels of boats that work in waters infested by marine borers.

Gurjun See *Keruing*.

Idigbo (*Terminalia ivorensis*). This is a light-weight hardwood, widely distributed in West Africa, yellowish in colour, with a fairly straight grain and rather coarse texture. Its density varies considerably averaging 540 kg/m³ (34 lb/cu ft). The core wood is liable to contain brittleheart with its associated compression failures. It is a stable and durable wood and therefore suitable for exterior joinery such as window frames. It reacts with, and is stained by, iron. It has been much used for the manufacture of plywood.

Iroko (*Chlorophora excelsa*). This tree, which is known as *mvule* in East Africa, grows right across central Africa from Sierra Leone to Mozambique. It produces a valuable and decorative wood which has many of the properties of teak but without the latter's smell and greasy feel, neither does it possess the rings of pores characteristic of teak. It seasons easily without much splitting or distortion. It has about the same density as teak—640 kg/m³ (40 lb/cu ft)—and the same hardness and shock resistance. It is a very durable timber and can be used for many of the exterior purposes for which teak was formerly employed. It is very suitable for window sills, draining boards, external doors, and for boat building.

Jarrah (*Eucalyptus marginata*). The timber from this tree, which grows only in Western Australia, was at one time imported in considerable amounts, but in this country it is not used much at the present time. It is a heavy, durable timber weighing about 800 kg/m³ (50 lb/cu ft), with correspondingly high strength properties. It is a rich dark red mahogany colour and it finishes well with wax, thus making a handsome floor. It has been used for heavy constructional dock work and was at one time made into sleepers for the London underground railways.

Kapur (*Dryobalanops spp*). This heavy, very durable timber comes from several species of *Dryobalanops*, a tree which grows to a large size in Malaysia. The light reddish-brown heartwood has a coarse but even texture and when freshly cut has an odour of camphor. Its average density is about 768 kg/m³ (48 lb/cu ft), and it is correspondingly strong, being about 50 per cent stiffer and tougher than teak. It is well suited for heavy constructional purposes.

Keruing (*Dipterocarpus spp.*) The timbers from various species of *Dipterocarpus* that grow in south-east Asia are marketed under names that denote the country of origin rather than the botanical species concerned. Thus the timbers from India, Ceylon and Burma are commonly called *gurjun*, those from Thailand, *yang*, those from Malaysia, *keruing* and those from the Philippines, *apitong*.

They are all coarse-grained woods varying in colour from light red to reddish-brown, and in density from 640–960 kg/m³ (40–60 lb/cu ft). They contain a resin which may exude from the surface of the sawn timber. They season slowly and are fairly durable. On the average they are tougher and rather stiffer than teak. They make useful timber for heavy engineering work and are good for flooring, and for window sills and other external joinery.

Mahogany. The name mahogany has been used for a number of somewhat similar timbers derived from trees belonging to the *Meliaceae* family. Originally it was applied to the handsome furniture wood which was shipped from Cuba, and hence called Cuban, or Spanish, mahogany. Supplies of this have long since been exhausted, its place being at first taken by a rather softer variety known as Honduras mahogany. This in its turn has now become scarce and most of our present day supplies come from trees in west Africa which yield rather similar timbers.

African mahogany is the timber of trees belonging to the genus *Khaya*, most of it coming from *K.ivorensis* which grows throughout the West-African forests. The pinkish brown heartwood of this tree may darken to a deeper reddish shade. The grain is often interlocked giving rise to a stripe or 'roe' figure when sawn on the quarter. There is considerable variation in the texture of the wood and the core sometimes contains brittle-heart and 'thundershakes'. The average density is 540 kg/m³ (34 lb/cu ft). The strength of the wood when free from brittle-heart is very similar to that of American mahogany though it is often harder. The outer heartwood is fairly durable but wood from the core is not and so should not be used for such jobs as planking of boats. Large amounts of African mahogany are used in the furniture industry, and also in the joinery trade for panelling and shop fittings. High-grade plywood is often made with veneers of this timber.

Makoré (*Mimusops heckelii*). This is a large West-African tree that yields a timber somewhat resembling a close grained mahogany. It is dark purplish in colour, rather denser than mahogany, averaging 624 kg/m³ (39 lb/cu ft) and therefore stronger in most aspects. It is also much more durable and, since it can easily be peeled, it is used for making plywood that is resistant to decay throughout its thickness.

Maple (Acer spp). The genus *Acer* includes a number of timber trees including rock maple, and sycamore (q.v.).

Rock maple (Acer saccharum). This tree in Canada is often called sugar maple as its sap is the source of maple syrup and sugar. It yields a non-durable, pale-coloured wood, hard and tough and of fine even texture, having strength properties superior to those of beech. Its average density is 720 kg/m³ (45 lb/cu ft). It is used where strength and resistance to wear are essential and is the timber of choice for flooring of dance halls, roller skating rinks and squash courts as it wears evenly and smoothly. It is also used for furniture and cabinet work. It turns well and is used in the textile trade for making rollers and in the shoe trade for the manufacture of lasts.

Soft maple. This timber, as the name implies, is not so hard as rock maple but, though inferior for flooring, it is a useful furniture wood.

Meranti (Shorea spp). The timbers shipped from Malaysia under the general name of meranti come from a number of related species of *Shorea*. They are classified for export on the basis of colour and weight, as light red, dark red, yellow, and white meranti respectively, or unsorted just as meranti. The average density of light red meranti is about 528 kg/m³ (33 lb/cu ft), but the density of the wood varies considerably according to the exact species from which it is derived. Other kinds of meranti are somewhat heavier and correspondingly stronger. They tend to have interlocked grain so that quartered material shows a striped figure. Their texture is moderately coarse but even. When of average density the timber is approximately equal to African mahogany in its strength properties. It generally works fairly easily and gives a good finish in most operations, but sometimes tends to tear when planed. The woollier grades need sharp, thin-edged tools to produce a smooth surface.

The merantis—with which may be classified red seraya

from Borneo—are suitable for joinery and general constructional work, and they can be used for exterior work as they are
fairly durable. They have also been used for plywood.

Oak (*Quercus spp*). The oaks fall into three groups which are
based on their wood structure:

> White oaks, which include European oak
> Red oaks, which are mainly American
> Evergreen oaks ('live oaks') of which the holm oak or ilex
> is a European example.

The possession of very wide rays is a feature common to all
oaks.

The deciduous oaks (both white and red) are ring porous,
generally with a well marked distinction between earlywood
and latewood.

European Oak is the timber of two closely related species of
Quercus, namely *Q.robur*, the pedunculate or common oak, and
Q.petreae, the sessile or durmast oak. The timber from these
two species cannot with certainty be distinguished and it
varies according to the region in which the trees have grown.
There are probably some varieties or races within these species
that include trees of better form than others and oak from some
regions has long had the reputation of being particularly
suitable for special purposes—e.g. Memel oak was highly
valued for barrel staves.

English oak has for centuries been regarded as possessing
to an outstanding degree the virtues expected of a hardwood,
viz. high strength and durability and a pleasing figure. It was
on account of having ample supplies of this timber that
England was able to build her famous wooden warships, such
as H.M.S. *Victory* which has been preserved until the present
day. And all the great half-timbered mansions that were built
in the days of the Tudors were constructed of English
oak.

The pale coloured outer sapwood zone is not durable and
is readily attacked by insects. The heartwood, on the other
hand, if kept dry, endures indefinitely and resists even the
death-watch beetle, unless the wood has been already softened
by decay. It is wise therefore either to remove all sapwood

from oak that is to be used in buildings or boats, or else to treat this vulnerable, permeable area thoroughly with a wood preservative before it is built into the structure.

In all the white oaks the vessels of the heartwood are blocked with tyloses and so the heartwood is very resistant to the entry and passage of liquids. This makes it very suitable for making casks to be used for beer, wines and spirits.

Oak seasons slowly and is very liable to split and check. It should therefore be piled with half-inch sticks and given end protection if conditions favour rapid drying. It is rarely kiln dried from green, a preliminary period of air drying being advisable. It has good bending qualities.

Oak is used for window sills and high class flooring and for a wide range of purposes where strength and durability are required, such as boatbuilding, dock work and fencing. It stains on contact with iron and therefore galvanised nails should always be used to fix it in situations exposed to moisture, for instance palings to fencing rails. On account of its acidic nature it can cause corrosion of metals and should never be used for packing-cases that may contain metal goods.

The attractive 'silver grain' in oak that results from the wide rays is displayed only when the timber is sawn on the quarter—i.e. more or less parallel to the rays. Decorative veneers cut from oak thus sawn are often used to cover chipboard and plain plywood to provide a decorative finish.

American white oaks. These are similar in many respects to the European and can be used for much the same purposes.

Red oaks. These trees yield a timber that is coarser than that of the white oaks and much less durable. As the vessels of the heartwood develop few tyloses the timber remains permeable and is therefore unsuitable for barrels. Turkey oak (*Q.cerris*) falls into this class yielding rather wide ringed wood with a wide sapwood band. Its timber is not durable.

Holm oak. Occasionally timber from the evergreen oak or ilex comes into sawmills. Its wood differs from ordinary oak being close grained without conspicuous annual rings. The hard dense wood is generally suitable only for rough work.

Obeche (Triplochiton scleroxylon). This rather featureless hardwood from West Africa is one of the lighter utility hardwoods having a density of only 380 kg/m³ (24 lb/cu ft. The pale yellowish wood shows no distinct heartwood. It is obtainable in wide boards free from defects but it is liable to stain if shipped in log form and generally requires quick extraction and treatment with anti-stain chemicals if this is to be avoided. The sapwood contains much starch and is susceptible to attack by *Lyctus* beetle. It is an easily worked timber which can replace softwoods for many purposes. It is used in furniture and joinery and for packing cases. It can easily be peeled and in Africa large volumes of plywood are made from it.

Poplar (Populus spp). There are several species of poplar that yield useful timbers, and a number of quick-growing hybrids raised through selective breeding have been planted extensively in many European countries. The woods from these varieties do not differ markedly. Most of the home grown timber comes from the black poplar.

Poplar is a light, soft, almond white timber weighing about 450 kg/m³ (28 lb/cu ft) when seasoned. There is little distinction between heartwood and sapwood. The wood has an even texture and is straight-grained but somewhat woolly. It seasons quickly and well, but it is not at all durable and logs should not be kept long on the ground after felling. It can be nailed as readily as a softwood and works easily provided that the tools used are kept very sharp. It is used for many purposes where a light, tough wood that does not splinter is required. It is particularly suitable for peeling into veneers for plywood and for making matches and chip-baskets.

Ramin (Gonostylus spp). Ramin is the standard name for the timber of several species of *Gonostylus* growing in Sarawak and Malaya. It is a pale, straw coloured, good quality, utility hardwood, weighing about 660 kg/m³ (41 lb/cu ft) and having strength properties similar to those of beech. It has a close even texture with no obvious heartwood. It is liable to bluestain and so must be converted quickly after felling and given anti-stain treatment to prevent bluestain developing during seasoning. It works cleanly and is a useful timber for furniture, interior joinery and turnery, but it is not durable enough for exterior use unless impregnated with a wood preservative.

Sapele (Entandrophragma cylindricum). This is a timber from West Africa of the mahogany type but appreciably harder and heavier than the general run of American or African mahoganies. It averages 624 kg/m³ (39 lb/cu ft). It has a pronounced stripe figure which shows best in quarter cut wood. From logs with wavy grain very decorative veneers with a fiddle-back mottle can be obtained. Its strength properties are similar to those of oak and it seasons fairly rapidly though with a tendency to distort. It is used for furniture, joinery, and shop fittings and it makes decorative flooring which wears well. Plywood is often faced with veneers of this wood.

Sycamore (Acer pseudoplatanus). Sycamore grows all over England, generally as single trees or in small groups. It has seldom been planted in large areas. The whitish wood resembles that of other maples having a fine close texture and strength properties very similar to those of oak, but it is not resistant to decay. In order to keep its clean white appearance the surface of freshly sawn planks must be dried off rather quickly, and narrow piling sticks should be used, or else the boards should be stacked on end to allow air to circulate freely over all the surfaces. This is a very popular wood in the turnery trade and is used for making mangle rollers, brush handles and bobbins. It is also very suitable for flooring. Some logs have a wavy grain and are cut into veneers which have a decorative figure.

Teak (Tectona grandis). This is unquestionably one of the most famous timbers in the world and is renowned for its great durability and its stability under changing conditions of moisture. Formerly it was imported from Burma and India but today supplies come mainly from Burma and Thailand. It is only moderately dense, weighing on the average 640 kg/m³ (40 lb/cu ft). It has a warm golden-brown colour sometimes showing darker markings, and it has an oily feel. Unlike many tropical timbers it is ring porous and shows distinct growth rings. Its strength properties are mostly superior to those of oak. It works easily but some operatives are allergic to its sawdust. It can be seasoned without degrade and moves very little with changes of moisture content.

Teak is the most useful timber in ship building, and on account of its low shrinkage and small movement is ideal for decking. It is also very suitable for deck houses, bulwarks, weather doors and hatches. Garden furniture, which is often exposed to all weathers, is usually made of teak and formerly the wood salvaged from old warships was often employed for this purpose. There has recently been a fashion for indoor furniture made from teak, but the timber is now so expensive that, as high durability is not necessary for this purpose, solid teak is seldom used and instead veneers of it are applied over a plain timber or chipboard.

Utile (Entandrophragma utile). This is very similar to the closely allied species sapele, but it lacks the cedar-like smell of the latter and is rather more open in texture. Also the stripe figure that is so characteristic of sapele is in utile wider and less regular. The reddish-brown heartwood is quite distinct from the paler sapwood. It is used for furniture, interior joinery and many of the same purposes as sapele. It can also be peeled to make decorative veneers for plywood.

Walnut (Juglans regia). European walnut has long been prized as a furniture wood on account of the interesting variations in colour and figure of the heartwood, and the ease with which it can be carved and finished. As supplies are now very short the best butts are usually cut into decorative veneers. It weighs about 640 kg/m³ (40 lb/cu ft) and is very stable when fully seasoned. It is the timber of choice for gun and rifle stocks.

American walnut comes from a related species of *Juglans*, but African and Australian so-called walnut come from quite different trees.

Willow (Salix spp). There are many species of willow, the most important of which in the United Kingdom is *Salix alba*. They grow best near streams and they are sometimes pollarded and yield withies for wicker work. The wood has a fine even texture similar to that of poplar. It is a low-density wood averaging about 450 kg/m³ (28 lb/cu ft). The sapwood is almost white and the heartwood pinkish.

A variety of *Salix alba* (var. *coerula*) has been found the ideal wood from which to manufacture cricket bats, being tough for its weight, and it is commonly called the cricket bat willow.

From lower qualities of this timber and from other species of willow artificial limbs, toys and chip-baskets are made.

Speciality Hardwoods

In the following section are briefly described a number of woods of which only relatively small quantities are available, but which have special, sometimes unique, properties that fit them for special uses.

Balsa (Ochroma lagopus). This is a very fast-growing tree of tropical America, the main supplies coming mostly from Ecuador. It yields the lightest timber in commercial use, the wood from young trees weighing only between 80 and 144 kg/m³ (5–9 lb/cu ft). Its clean wood has a whitish or oatmeal colour, but it is rather prone to fungal staining if there is delay in conversion and seasoning. It is soft and easily in- dented but rather tougher than one would expect from its very low density. Very sharp, thin-edged tools are required to get a clean finish when working with it. It has a very low thermal conductivity and before the invention of expanded polystyrene was extensively used for insulation. Today its principal use is for making model aircraft. During the war large quantities were used in the sandwich construction of the Mosquito aeroplane.

Boxwood (Buxus sempervirens). In this country the box seldom grows to a size worth converting into timber and our supplies come mainly from the Near East. Small logs are imported in 4 ft lengths, 4–8 in in diameter. The light yellow wood is extremely dense with a very fine texture and weighs on the average about 900 kg/m³ (57 lb/cu ft). Great care must be taken to season it slowly to avoid splitting and degrade. It is used for rulers and for the turned handles of tools that have to stand up to much hammering. It is very suitable for turning and for carving small figures such as chessmen.

Other similar timbers which are used as substitutes for European boxwood include East London and Knysna (from South Africa) and Venezuelan boxwood.

Cocuswood (Brya ebenus). This wood, which comes from the West Indies, is very heavy, dark chocolate brown in colour and beautifully veined and is used for the making of wooden wind instruments. Its fine, uniform, dense texture and oily nature make it an ideal wood for the finest turnery.

Ebony (Diospyros spp). Ebony comes from the heartwood of a number of species of *Diospyros* which grows in West Africa and India. In the case of ebony from East India great care must be taken in seasoning and it should be cut into small sizes before drying. The best ebony is almost black with a very fine even texture. Owing to its hardness it is difficult to work but it turns well and takes a splendid polish. It has long been used as a decorative wood for inlay work, and it is also used for handles and for the small parts of stringed instruments.

Hickory (Carya spp). Hickory comes from several species of *Carya* which grow in eastern North America. It is a hard, heavy wood which is exceptionally tough and very stiff being stronger in these respects even than ash. It bends well and it is probably the best timber in the world for the handles of striking tools and for shunting poles. Formerly it was much used for the shafts of golf clubs.

Lance wood (Oxandra lanceolata). This very hard, heavy wood comes from the West Indies. It is of a pale yellowish colour and is noted for its strength and resilience which make it an ideal wood for archery bows and for the top joints of fishing rods. It has also been used to make meat skewers and small handles.

Lignum vitae (Guaiacum spp). This is one of the hardest and heaviest woods known, weighing on average over 1200 kg/m³ (about 77 lb cu. ft). It comes from the West Indies. Its resistance to indentation exceeds that of almost all other woods and it is extremely hard to work. It is very durable and almost immune to termite attack. The heartwood is a very dark greenish brown and from it a gum can be extracted which was formerly used in medicine.

Owing to its hardness and oily nature this wood is the ideal material for making the bushes and bearings of propeller shafts for ships. The 'woods' used in the game of bowls are also made from this timber.

Muhuhu (Brachylaena hutchinsii). This is a dark yellowish-brown wood from East Africa which makes excellent heavy duty flooring being very hard and having a fine, even texture which resists abrasive action.

Muninga (Pterocarpus angolensis). Muninga is a very attractive wood from central Africa. The colour is a rich dark brown becoming golden on exposure. Owing to its curly grain it often exhibits an interesting figure and is used for panelling, furniture and high-class joinery work.

Persimmon (*Diospyros virginiana*). This wood comes from a species of ebony but as it forms little heartwood it has only a very pale brown colour. It is a hard heavy wood with a close even texture. It has mostly been used for the heads of golf clubs and for making shuttles.

Rosewood (*Dalbergia spp*). Rosewood is a handsome, heavy timber which comes from India. It has a dark purplish colour and a blackish streak at the end of each growth zone which gives rise to an attractive figure when sawn. It is rather hard to work but turns well and takes a high polish. It has long been used for the highest class of cabinet work and furniture. Usually today veneers of it are applied to a plainer timber or chipboard.

Honduras rosewood is rather harder and stronger than that from India and has been used for decorative handles.

Satinwood (*Chloroxylon swietenia*). This wood comes from a smallish tree that grows in Ceylon and India. It is a very hard timber of beautiful lustrous golden yellow and has long been used for fine cabinet work and for making fancy goods, sometimes being applied as a veneer to less valuable woods.

BIBLIOGRAPHY

Forest Products Research Laboratory. *Handbook of Hardwoods* 2nd edition. (London, H.M.S.O.), 1973.

CHAPTER ELEVEN

Major Uses of Wood

Wood is a most versatile material and from time immemorial has been used for a great variety of purposes. In the past it was often the only material that could be worked with the tools then available. In many countries, where forests were being cleared for agriculture, it was so cheap that nothing else was competitive in price. Today the situation has changed. Timber is no longer cheap, and technological discoveries have created new materials which have some, if not all, of the qualities of wood without suffering from its less desirable features, such as swelling and shrinking on wetting and drying. Steel has replaced timber for the structural units of larger buildings, concrete sleepers have replaced wooden ones on main line tracks, and fibreglass is now widely used to make boats.

If timber is to maintain its present uses in competition with these other materials it is essential not only to choose the right kind of timber for the job but also the correct grade of that timber, and then to ensure that it is seasoned to the right moisture content. If there is any risk of the timber becoming damp in service, or there is a chance of its being attacked by insects, a durable species, or one that is permeable to wood preservative treatment, should be chosen. Many markets for timber have been lost, wholly or in part, because adequate precautionary measures against decay have not been taken. For example, failure to treat external window joinery properly with preservative has led to the partial replacement of woodwork by metal windows, and for the same reason wall cladding (siding) is now often made of white plastics.

When designing a structure and choosing the materials to be used for it the following qualities must be considered: strength, stability, availability in the required sizes, ease with which the materials can be worked, durability, the amount of decoration and treatment that will be required to maintain it, thermal insulation, fire resistance, and appearance.

As already explained the sources of supply of timber are continually changing and the user may pay far more than is necessary if he insists on buying a timber that is becoming increasingly scarce instead of equally good, and much more widely available, alternatives. It is also, of course, wasteful to use a higher grade than is necessary. Similar timbers can often be substituted for the traditional ones without loss of efficiency. European whitewood, for example, is just as good as European redwood for flooring, although for some other purposes it may be quite unsuitable. For instance it cannot be substituted for redwood for fencing that is to be creosoted, as it is not permeable to preservative liquids.

A timber-merchant should find out the purpose for which a customer requires the wood he is ordering and, if the latter is not fully informed about the quality and grade of wood he needs, should suggest the one best suited for the job. Goodwill may be permanently lost if a merchant sells a parcel of timber that turns out to be unsuitable.

Suggestions are given below as to timbers suitable for particular purposes at the present time, but others equally suitable may well become available in the future as market conditions change.

In Europe softwoods are generally preferred to hardwoods for most structural purposes, such as carcassing timbers, rafters and joists, because they are cheaper, easier to work, and are usually available in a wide range of widths and lengths. Many species of softwood are more or less interchangeable for general building work.

Hardwoods are much more variable and so greater care

must be taken to select the right species for the job and to ensure that it is properly seasoned to the required moisture content.

Building Timber

European redwood and whitewood, Douglas fir, Western hemlock and home-grown Sitka spruce, are the softwoods most commonly used for building timbers in the United Kingdom; but radiata pine, after preservative treatment, is succesfully used in those countries where it has been planted on a large scale.

Boats

The traditional timbers for building ships and large boats are oak, elm and teak. European oak is probably still the best all-round timber for frames, stringers, keels and deadwoods. Rock elm, which bends easily, has long been regarded as an ideal wood for the frames. It is not, however, very durable, and the other American elms are even less so. English elm has been used for the deadwood and hog, and for below-water planking, but this too is not very durable unless permanently submerged under water. Teak is now so expensive that it can rarely be afforded for anything except the highest class of work, but afrormosia makes a good substitute.

Planking is most often made of African mahogany, but if this is used for hull planking it should be of a good dense quality cut from the outer heartwood (not the core) of the logs. Larch, if straight grained and good quality, also makes excellent and durable planking.

Racing eights and skiffs are often made of Canadian Sitka spruce.

For the decking of boats teak is the ideal wood but afrormosia and agba are other hardwoods that have been used successfully for this purpose. Rift sawn Douglas fir and pitch pine are also suitable. The timbers used in the boatbuilding industry in 1964 were listed by Thomas (see bibliography).

Casks

Casks are of two kinds—those made to contain liquids and those intended for solids. The former are known as 'wet' or 'tight' cooperage, and the latter as 'dry' or 'slack'.

Wet cooperage. Beer casks were formerly made of oak, either European or American white, but today kegs made of aluminium or stainless steel have largely taken their place on account of the greater ease with which they can be sterilised. Whisky, however, must by law mature for at least three years in wooden casks or vats. American white oak has for a long time been the favourite timber for these, but red oak is quite unsuitable being much too porous. Oak sherry butts are often reused for the storage of spirits. Wine and fruit juices are often imported in casks made from sweet chestnut and these are often used again for packing crockery and as flower tubs.

Dry cooperage. This trade has diminished considerably since the beginning of the century but there is still a demand for casks to contain such things as early potatoes, herrings, and various other food-stuffs. Light-coloured softwoods without appreciable odour, such as spruce, are preferred for this purpose.

Dock and Harbour Works and Marine Piling

Durable timbers in long lengths are required for dock gates, marine piling etc. and so the choice is limited. Greenheart has long been the favourite wood for these purposes but other tropical woods have lately been used such as opepe, pyinkado, iroko and wallaba. Oak is probably the best home-grown wood for the purpose, while among softwoods long-leaf pitch pine and Douglas fir, creosoted after incision, are suitable.

Flooring

Broadly speaking flooring is of two types: that which is exposed to all the wear and that which is covered with carpets, linoleum, or other such material.

Timbers suitable for the first type must be reasonably hard and wear evenly without splintering. They must take a good finish without becoming dangerously slippery, and they should have a pleasing appearance. Generally speaking it is among the hard woods that timbers with these qualities are found. When choosing which to use consideration must be given to the type of wear to which the floor will be subjected. Heavy industrial use, hospitals, schools and dance floors etc will obviously require a wood harder and more resistant to abrasion than normal domestic usage demands.

Timbers suitable for domestic floors include the following:

Oak   — preferably free of sapwood, or else with the sapwood treated with a preservative.

Beech — particular care must be taken to season this to the correct moisture content and to protect it against damp in use.

Maple— perhaps the best flooring wood available.

Though the above are probably the best, the majority of hardwoods whose density exceeds about 640 kg/m³ (40 lb/cu ft) can make good flooring and the following have been used successfully: afrormosia, birch, danta, gurjun, iroko, jarrah, muninga, opepe, purpleheart, sapele, teak and yang.

For floors that will receive much heavier wear, woods exceeding about 800 kg/m³ (50 lb/cu ft) are generally required, e.g.: East African olive, greenheart, muhuhu, okan, pyinkado, Rhodesian teak and wallaba.

(These can of course also be used for less exacting conditions.)

For floors that are to be carpeted or otherwise covered, and are to be exposed only to light wear, softwoods are generally used. European redwood or whitewood, Western hemlock, and rift-sawn Douglas fir are all suitable.

Furniture

Woods for the manufacture of furniture can be divided into those that provide the structural framework to which upholstery etc. is to be fixed; and those that are to be used for the surfaces that are visible and which therefore must have an attractive appearance.

Beech has for a long time been the timber of choice for the structural members of chairs, but ramin in recent years has also been used satisfactorily for this purpose.

Oak has been used for many centuries to make tables, chairs, chests, ecclesiastical furniture and so on. Limed oak had a fashion in the interwar years. Elm has been the traditional timber for the Windsor chair and this wood is used very largely by one of the leading furniture manufacturers of this country.

In the nineteenth century mahogany from the West Indies was the favourite furniture timber and African mahoganies are now used extensively both for modern designs and in reproductions of Regency-style furniture.

Walnut was much in vogue in the eighteenth century but today is so valuable that it is normally used only in the form of veneers over a plain timber or chipboard.

Antique pieces of furniture made of exotic timbers such as satinwood, rosewood and ebony are now collectors' pieces.

Fashion exerts a powerful influence over the design of furniture and the materials of which it is made. Some years ago light-coloured timbers such as birch and ash were all the rage. The present fashion is for furniture finished with a teak veneer. What will replace this is anyone's guess but there is a wide range of handsome timbers abailable from which designers can choose. The timbers and board materials used in the furniture industry have been listed by Webster, 1966.

Fencing

The traditional timbers for fencing posts and palings were oak and larch because these were the most durable available in country districts. Today, when plants for impregnating timbers with creosote or copper-chrome-arsenic exist in most areas, many other kinds of wood can be used, provided they are permeable. For instance beech, birch and pine, if properly impregnated, make excellent and durable fencing posts. Round poles from thinnings are particularly suitable for impregnation as they consist almost entirely of sapwood which is very permeable and readily absorbs preservatives. Sweet chestnut poles are split to make the cheap wire fences often used for temporary purposes.

Greenhouses

Although aluminium alloys have to a great extent replaced timber for the larger commercial greenhouses and frames, many wooden ones are still made for the amateur gardener. Western red cedar is the timber most commonly used as it is both durable and easily worked and does not require painting, but oak is also used quite frequently. The life of cedar and oak greenhouses and frames can be greatly extended if they are occasionally given a brush-applied treatment with a solvent preservative—creosote should never be applied to timber near living plants. Western red cedar is also the ideal timber from which to make Dutch lights, beehives and seedboxes.

Joinery

Timbers for joinery need to be straight grained and free from large knots. They should also be stable and move little with changes in moisture content, and they should be reasonably easy to work. A distinction must be made between those that are suitable for external work and those that should be used only indoors.

For external use, only timbers that are at least moderately durable or can readily be treated with a wood preservative should be used. Formerly, good quality European redwood from northern Sweden and Russia was the principal timber for window frames. This was selected so as to exclude almost all the permeable and non-durable sapwood. But after the Second World War this was no longer available and joinery was made from timber containing a fairly large proportion of sapwood. This resulted in extensive decay of windows and doors occurring within a few years of the houses being completed. It is now recognised that external joinery made of non-durable timbers must be treated with a wood preservative (see p. 91) before being painted.

Though it has always been considered good practice to make window and door sills from a durable hardwood, such as oak (heartwood) or teak, there are several other suitable woods now available including gurjun, iroko, idigbo and opepe.

For internal joinery there is a much wider choice of timbers. Softwoods are generally used for joinery that is to be painted, but usually a decorative hardwood is chosen for panelling that is to be varnished or wax polished, and the choice is often made on the basis of colour and figure. If a light-coloured wood is required the choice may be made from ash, agba, oak, obeche, ramin or sycamore; but if a darker wood is preferred a dark red African mahogany, sapele, or African walnut may be selected.

The requirements for quality of timber in manufactured joinery are specified in BS 1186 of 1971.

Packaging

Very large quantities of timber are used annually to make boxes and packing cases. The requirements of wood to be used for this purpose are:

(i) It should be strong enough to protect the contents of the case or crate.

(ii) It should be able to take nails without splitting.

(iii) It should have no deleterious effect on the contents—e.g. no corrosive action on metals, or no strong smell that might affect contents susceptible to tainting, such as foodstuffs or tobacco.

Softwoods from pines and spruces are generally used for case making but poplar is also suitable. The grade and quality of the wood will, of course, depend on how strong the case is required to be. Heavy machinery, for instance, will require a very strong case made of good quality timber and free from knots and cross grain.

Pallets

These are boarded platforms raised some inches off the ground on bearers or blocks. They are being used increasingly for the rapid handling of goods by fork-lift trucks. The boards for these are generally of softwood (e.g. Maritime pine) while the blocks or bearers are usually made from home-grown hardwood such as elm.

Pitwood

Much less timber is used in mines today than was formerly the case as the roof of the main workings is now generally supported on steel arches with some packing of wood above them. Large numbers of pit-props are however still required and thinnings from softwood plantations are generally used for them. They must be straight, free from large knots, and well prepared. If pretreatment with a wood preservative or fire retardant solution is required they must be debarked and air dried down to about 25–30 per cent moisture content before impregnation.

Plywood

The choice of timbers for veneers to be made into plywood is discussed in Chapter 12.

Poles

European redwood is the preferred timber in the U.K. for poles to carry telephone wires and for electrical transmission lines. It has a wide ring of sapwood which can readily be impregnated with creosote and this ensures a long life. It also has adequate strength and is not too heavy to handle conveniently. Larch and Douglas fir have also been used but they require longer impregnation treatment to achieve sufficient penetration of preservative.

Pulp

It is possible to make pulp from almost any kind of wood, but hard, heavy woods and those that contain large amounts of resin or other extractives are more difficult to process and produce pulp of inferior quality. The most suitable timbers for making good quality paper are pale-coloured softwoods such as spruce and silver fir.

Railway Sleepers

Although most of the sleepers on main line tracks in Great Britain are now made of reinforced concrete, large numbers of wooden sleepers are still required for replacements in secondary lines. European redwood (Scots pine) with its broad sapwood that can readily be impregnated with creosote has long been the timber of choice in the U.K. but in parts of Europe, notably Denmark and France, beech, which also absorbs creosote well, has been used extensively. In countries where durable hardwoods such as jarrah or iroko are available locally they have been used for this purpose, but they are generally too expensive if they have to be imported.

Sports Goods and Tool Handles

For tennis rackets, hockey sticks and the handles of striking tools, toughness—i.e. high degree of resistance to suddenly applied stresses—is the supreme requirement. Ash and hickory are the two most notable woods in this respect. Ash for these purposes should be moderately fast grown with a high proportion of dense latewood in the annual rings. Willow is, of course, the only wood accepted for making cricket bats.

Turnery

The shaping of a piece of wood by revolving it against a
stationery cutting tool is an ancient craft still practised today.
Broadly speaking turnery is of two kinds—firstly what might
be called 'utility turnery', i.e. such goods as broom heads,
brush handles, cheap toys, etc. much of which trade has now
been lost to plastics; and secondly ornamental turnery producing handsome articles such as fruit-bowls, table-lamp
stands and the like.

The essential requirement of wood for turning is the ease
with which it can be cut on a lathe to give a clean finish on
both the radial and tangential faces. Woods that have a fine
even texture are therefore most suitable. The best for utility
turnery include alder, beech, birch, cherry, lime and horse
chestnut; while for ornamental work mahogany, teak, and
walnut are ones most often used. Smaller fancy articles are
made from valuable timbers such as box, olive, rosewood,
ebony and satinwood.

Vats

Until recent times wood was the best material for making
vessels that had to contain solutions corrosive to mild steel or
iron. Fermentation vessels in breweries were all made of timber
as were all the storage vessels for wine. Nowadays, however,
stainless steel is frequently used for these purposes. Kauri pine
from New Zealand was the timber of choice for vats in the
past as it was free from knots, durable and perfectly straight
grained. Today when wooden vats are required other timbers
have to be considered as kauri pine of that quality is no longer
obtainable. There are quite a number of suitable woods, including prime grades of pitch pine, Douglas fir and larch.
White peroba from South America is one of the most successful
of the tropical hardwoods but is not always obtainable. Where
resistance to chemical attack is required purpleheart has given
very good results, but it can only be used for certain purposes
as it may release some red colour. Vats for spirits are almost
invariably made of oak.

BIBLIOGRAPHY
Tack, C. H. *Joinery*. (London, H.M.S.O.), 1971.
Thomas, A. V. *Timbers used in the Boatbuilding Industry*. (London, H.M.S.O.), 1964.
Webster, C. *Timbers and Board Materials used in the Furniture Industry*. (London, H.M.S.O.), 1966.

CHAPTER TWELVE

Composite Wood

As explained in Chapter 16 it is likely that in the future an ever increasing proportion of the wood from the forests of the world will be converted into some kind of composite wood, and a lesser amount will be used in the form of solid timber sawn from the log. The reasons for this are partly technical and partly economic and are as follows:

  (i) A manufactured wood product can be made to a constant composition having a specified uniform strength per unit area or volume.
 (ii) Board materials can be made to any required width, thickness and density.
(iii) The anisotropic (directional) strength properties of solid timber can be changed to the more uniform strength properties of the composite wood products, and the influence of growth features and defects on strength can greatly be lessened.
 (iv) Attractive decorative finishes can be applied to one or both faces of board materials during manufacture so that no decoration of the board is required subsequently.
  (v) Board materials can be made from young trees grown on short forestry rotations and from timber of relatively poor quality.

Composite wood products are basically of two types: those that consist of layers of wood glued together to form a solid piece of much greater thickness than the individual layers, e.g. laminated beams and plywood; and products made from wood disintegrated mechanically into chips or shavings and then bonded together under high pressure after the addition of a small amount of a synthetic resin adhesive, e.g. chipboard.

Adhesives for Timber

When the only glues available were those based on milk and other proteins derived from animals, laminated products could only be ssed safely in dry conditions. Otherwise the glues, being susceptible to damp and microorganisms, decay and the product falls apart. It was only with the development in the 1930s of adhesives based on synthetic resins, which were not affected by these hazards, that the field for composite wood products was widened to include service conditions involving full exposure to the weather.

As a result of a long series of exposure trials in a variety of situations at the Forest Products Research Laboratory the relative durability of a wide range of commercial adhesives has been assessed and the most suitable types to use for many purposes can now be specified.

A classification of glues on the basis of their durability has been laid down in BS 1203 and four types are described as follows:

Type WBP   Weather proof and boil proof adhesives. Phenol formaldehyde resorcinol resins. These make joints highly resistant to weather, microorganisms, boiling water and dry heat.

Type BR   Boil resistant. Mainly melamine resins. Joints made with these have a good resistance to weather but ultimately fail under very long exposures, which WBP would be able to withstand. They are very resistant to cold water soaking and to microorganisms.

Type MR   Moisture resistant and moderately weather
               resistant. Based on urea formaldehyde. They are
               resistant to microorganisms and withstand wet
               conditions, but will only survive full exposure to
               weather for a few years.
Type INT   These adhesives include casein, animal glues,
               and extended urea resins. They have some
               resistance to moisture but do not resist microbial
               attack. They last indefinitely in dry conditions.

Adhesives can also be classified on their ability to fill gaps
and slight irregularities between wooden joints. Gap filling
adhesives complying with BS 1204 must be able satisfactorily
to fill gaps up to 1·27 mm (0·05 ins) without crazing. These are
used in joints where the clamping pressure cannot ensure a
tight fit. Close contact adhesives are used in accurately
machined woodworking joints, and where adequate and even
pressure to hold the surfaces in contact can be applied during
the curing (setting) process. The rate of curing of synthetic
resin adhesives is greatly accelerated by exposing the joints
to an elevated temperature.

Considerable research has gone into preparing glues that
have precisely the requirements of particular industries, and
many special formulations are now available commercially.
But an adhesive can only function properly if the wood to
which it is applied is at a suitable moisture content, and if it is
applied at the correct rate of spread. The moisture content of
the wood should not exceed 15 per cent and should be within
3 per cent of the mean equilibrium moisture content of the
assembled timber in service.

Plywood

The first composite wood material to be manufactured on a
large scale was plywood. This has been defined as a product of
balanced construction made up of plies assembled by gluing,
the chief characteristic being the crossing of the grain in
alternate plies to improve the strength properties and to
minimise movement in the plane of the board.

The production and consumption of plywood has increased ever since the beginning of this century. At first much of it was made cheaply, largely for the construction of tea chests; but with the development of water resistant resin glues the usefulness of plywood, and the scope for its use in a great variety of situations, has greatly increased the demand. In the ten years between 1960 and 1970 the imports of plywood into the U.K. increased from 449 100 to 821 600 m³. Canada and Finland together supplied more than half of this amount, with the Soviet Union sending over 108 000 m³.

The veneers that constitute the plies are made by peeling a round log, or by slicing from flitches or partly squared logs. The latter method is used only when it is wished to show the grain in a particular way.

Logs selected for peeling are debarked and then softened by steaming, or steeping in hot water. They are then cross cut to the length required and clamped in a huge lathe which turns the log against the peeling knife. Considerable lengths of almost continuous veneer are thus produced and are wound onto a spindle. This is transferred to a conveyor where the veneer is unwound and passed through a guillotine which cuts it into the size of the press in which the plywood will be made, a common size being 8 × 4 ft. The veneers are next dried down to a moisture content of 6–8 per cent and graded three qualities, face, back and core. From the face veneers knots may be punched out and replaced with a clean piece. Any narrow strips left after defects have been trimmed away are joined by their edges to make sheets of correct width. The dry sheets of veneer are now coated with adhesive by passing them through a glue spreading machine and then put together in their correct sequence and direction of grain. Each layer has the grain running at right angles to those adjoining it. The assembled plies are finally stacked between metal sheets in a large heated hydraulic press and, after pressing, trimmed to the exact size required and packed for dispatch.

Many kinds of wood can be peeled to make satisfactory plywood, the most suitable being those with a close, even texture. The traditional timber for this purpose in northern Europe has been birch, and in North America, Douglas fir, both of which yield a strong plywood with a specific gravity of over 0·5. Supplies of birch in northern Europe being insufficient to meet the demand much plywood from the Baltic is now being made with a spruce core. Good plywood can be made from such European woods as beech and poplar, and considerable quantities are now being manufactured from gaboon, afara and obeche in Africa, and from luan and similar woods in Malaysia. Durable timbers such as makoré can be used to make plywood for special purposes such as boat building. Decorative veneers from ornamental timbers sliced from flitches of woods such as walnut and sapele are generally kept in the order that they have been cut so that adjoining veneers can be matched to give a symmetrical pattern.

Plywoods are made in a number of thicknesses varying from 3–25 mm, but that from Douglas fir is made from $\frac{1}{4}$–1 in thick. Thin plywood is generally made from three veneers. Multi-ply boards have an uneven number of plies so that the grain on the face and on the back always run in the same direction.

Assuming the strength of the bond between the veneers is as great, or greater, than the cohesion of the wood fibres between themselves, the strength properties of plywood are closely related to those of the timber, or timbers, from which it is made. But whereas defects in the wood rarely have more than 10 per cent effect on any strength property of plywood because of their dispersion throughout its thickness, in solid timber knots or cross grain can often render it unfit for any use where strength is important.

Table 5 reproduced from *Forest Products Research Bulletin* No. 42, gives the comparative strength properties of dry plywood made from various timbers. The excellent strength properties of birch plywood may be noted.

Plywood is chosen in place of solid timber when dimensional stability, rigidity, and good resistance to splitting and checking are particularly important. Its availability in wide sizes makes it especially advantageous as sheathing for partitions, walls and roofs, and very large quantities are used as shuttering for concrete. Over a quarter of the total consumption of plywood is used for joinery and furniture, and it is also particularly suitable for containers where its low weight and stiffness give it distinct advantages over metal.

TABLE 5. Basic stresses for plywood. Dry exposure

| Species | Modulus of elasticity N/mm² | | Fibre stress in bending N/mm² | Compression N/mm² | | Tension N/mm² | | Panel shear N/mm² |
	Parallel to grain	Perpendicular to grain		Parallel to grain	Perpendicular to grain	Parallel to grain	Perpendicular to grain	
Afara	11 380	340	22·8	21·4	0·28	24·1	0·48	2·41
Agba	8 960	210	15·9	14·5	0·62	22·8	1·10	2·21
Canadian birch	13 450	760	29·0	27·6	2·28	51·7	4·34	6·07
Daniellia	7 000	140	10·3	9·0	0·14	11·7	0·21	1·72
Danta	12 760	210	26·2	22·8	0·55	30·3	0·76	2·96
African mahogany	8 960	280	15·2	15·2	0·28	19·3	0·41	2·62
Makoré	12 760	480	26·2	26·2	1·10	27·6	1·10	3·24
Niangon	12 070	410	22·8	22·1	0·76	24·1	0·83	3·24
Ramin	16 200	340	33·1	32·4	0·90	37·2	1·03	3·52
Sepetir	12 070	620	31·0	29·6	1·31	35·9	1·45	3·59
Red seraya	10 000	210	15·2	13·8	0·21	24·1	0·34	2·00
White seraya	11 720	210	15·9	14·5	0·28	24·8	0·69	2·21
'Chile pine'	12 410	210	19·3	17·2	0·28	25·5	0·41	2·76
Scots pine	8 960	210	16·5	15·9	0·55	24·1	0·76	2·76

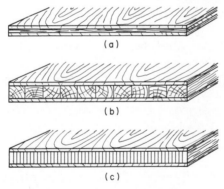

(a)

(b)

(c)

29. Composite boards (a) plywood
(b) blockboard
(c) laminboard

Plywood can be impregnated with standard wood preservatives or fire retardant solutions after manufacture, or the veneers can be treated with a water-borne wood preservative before gluing and assembling.

Plywood faced with sheets of decorative veneers, or plastic reproductions of wood grain, are now available. Sometimes one face is covered with thin sheets of metal or with a surface film impregnated with synthetic resin, and this finds many uses where a durable moisture resistant material is required for external use.

Grading of plywood is done according to the appearance and finish of the exposed faces, but the grading rules generally apply only to the product made in the particular country of origin. British manufacturers produce plywood to BS 1455, 1963. The grading rules are quoted in the T.R.A.D.A. bulletin on plywood.

Blockboard

Blockboard resembles plywood in that it has veneers on both sides, but the core consists of strips of solid wood up to 2·5 cm wide glued together side by side.

Battenboard is a similar material but the strips in the core are about 7·5 cm wide.

Laminboard has a core of thin strips glued face to face.

These materials, like plywood, are often faced with veneers of mahogany or other decorative hardwood. They provide strong rigid sheet material which is very suitable for doors, counter and table tops and shop fittings.

Finland provides most of the blockboard used in the U.K. supplying, in 1970, 162 900 m³ out of a total import of 203 900 m³.

Laminated Products

Laminated timbers are made by gluing together boards of seasoned timber with the grain running parallel in each layer. By this means pieces of very large cross-section and great length can be built up, and the weakening influence of features such as knots is dispersed. It is also possible in this way to produce timbers of great size uniformly seasoned throughout. Again it is possible by previously impregnating the laminae

30. Applying adhesive to boards being assembled to form a large laminated
member. (C.I.B.A.–Geigy)

with a wood preservative to produce beams treated right
through with preservative. Such complete treatment of
large-sized timbers cannot possibly be achieved for solid
timber of comparable dimensions. Further advantages are
that the thickness of structural members may be increased at
the points of maximum stress (see Fig. 31), and curved parts
can be made to a much smaller radius than would be possible
with solid wood of the same cross section. Hockey and shinty
sticks, and tennis rackets, all formerly made by bending solid
timber, can now be made with much less difficulty and waste
from laminations about 4 mm thick. Golf club heads which
formerly were made of persimmon, are now usually made of
laminated maple protected with an epoxy resin.

Lamination has made possible a whole new technology of
timber engineering. So-called 'glulam construction' permits

31. Building up laminated iroko stem of large yacht. (C.I.B.A.—Geigy)

the engineer to design shapes and sizes that cannot be achieved with solid timber, and many of these structures with their graceful curves are aesthetically very pleasing. Many sports stadia, churches, assembly halls, and such structures as bridges that require wide unsupported spans have been built with glulam construction. Laminated timber is also particularly suitable for situations in which severe corrosion of metals is

32. Laminated knees being bolted in ketch. (C.I.B.A.—Geigy)

likely to occur, as in certain chemical factories and dye works.

Laminated timber can be stress graded and three such grades are given in BS CP 112, 1967. At the Princes Risborough Laboratory a machine has been tested for converting in one operation random length planed boards with serrated ends into a continuous laminated beam, which can then be cross-cut to any required length. These serrated ends are made to fit into each other after being coated with glue—a process known as finger jointing. An authoritative assessment of this process was published by Sunley (1969).

This has made it possible to convert timber sawn from small trees into standard sized laminated members for use in buildings. The individual laminations are usually between 32 and 44 mm, but for curved members rather thinner boards, between 13 and 25 mm, which can bend more easily, are used.

The timber laminating industry has grown steadily during recent years as architects and designers have come to appreciate the many benefits of laminated construction.

Particle Boards

The term 'particle board' covers all panel material manufactured under pressure from particles of wood, e.g. wood
chips or sawdust, with or without the addition of an adhesive.
The commonest form is *chipboard* which is made by binding
wood shavings or chips under high pressure with a small
amount of a urea formaldehyde adhesive. It was developed
during the 1940s and its production and consumption has
increased remarkably in the last few years. The consumption
of this type of board in the U.K. rose from 304 000 tons in
1969 to 420 000 tons in 1971. Production has also increased
greatly in Great Britain during the past decade, and a very
large new plant started production in Scotland in 1973.

At first the production of chipboard was regarded as a
convenient way of using up factory waste wood and forest
thinnings, but it has now become an important industry in its
own right, chipboard having been found to have many
advantages over solid wood. Some of these are its consistent
quality, its freedom from knots, and the large sizes of sheets
that can be supplied, e.g. $7 \cdot 5 \times 2 \cdot 5$ m.

In the past softwoods have been the main source of raw
material for chipboard manufacture in the U.K., but recent
research at Princes Risborough has shown that satisfactory
boards can be made from hardwoods such as khaya, oak and
beech and from mixtures of hardwood with Scots pine. It is
made in various thicknesses, ranging from 9 to 25 mm and
with a density of 550–650 kg/m^3. Higher density boards are
made for flooring for which purpose chipboard is being used
to an increasing extent as it is some 25 per cent cheaper in
labour costs to lay than boards. When manufactured in the
U.K. it is marked 'Flooring Grade'. Large quantities are also
used in the furniture industry as it forms a stable core over
which veneers of wood or melamine laminates can be applied.
By using a much finer grade of chips for the surface layers it is
possible to produce chipboard with a surface almost as smooth
as hardboard (see below). The weather resistance of present
day chipboard is generally considered to be too low to permit
its use for exterior work.

Fibre Building Boards

Fibre boards are manufactured from wood by a process that
is basically similar to that employed in making paper or
cardboard. Three main types are made:

Insulating boards which are soft boards of low density e.g.
240 kg/m³. They are used for lining walls, ceilings and roofs
to prevent heat loss. (Some are made from cellulosic
materials other than wood such as sugar cane bagasse).

Panel boards which are medium density boards used for
partitions and wall linings.

Hardboards which include *standard hardboard* and *tempered
hardboard*. These are both of high density—up to 960 kg/m³.
Standard is used for floor underlays, kitchen units and
flush doors. The tempered hardboard has a drying oil
incorporated in it during manufacture. This gives it an
enhanced resistance to moisture so that it can be used for
concrete shuttering and external cladding. Recent studies
at Princes Risborough indicate that this material can also
be used for structural purposes in much the same way as
plywood.

A good survey of the structure applications and information
about the types, sizes, and proprietary makes of fibre board
was published in the magazine *Wood* in 1970.

Improved Wood

An intimate association between wood and plastics, known as
'improved wood', can be achieved by compressing at a fairly
high temperature thin veneers of beech interleaved with a
resin impregnated paper, by impregnating veneers of beech-
wood with a resin in solution and curing the pack under high
pressure at an elevated temperature, or by impregnating a
permeable wood with a vinyl monomer and then polymeri-
sing the latter within the wood.

By any of these means extremely strong, stable materials,
with a density over 1300 kg/m³, can be produced. Such

materials can be machined to a precise finish and are used for insulating components in electrical installations. They have also been used for tool handles, and for bobbins and picker sticks in the textile indutry. Densified laminated wood can be turned and made into cog wheels, nuts and bolts, and so on and, as it has a high resistance to corrosion, may take the place of metals where the risk of corrosion is high.

Wood Wool Cement

Wood wool, which is made by scraping long strands of wood from small softwood logs, is commonly used for packing fragile materials such as ceramics. It can also be coated with liquid cement and pressed to make large wood wool cement blocks. This material has very good insulating properties and is resistant to decay and to fire. It is therefore very suitable for making internal partitions and for insulating roofs.

A similar type of material called 'Strandboard', in which the wood is bonded together with a urea formaldehyde glue, was developed at the T.R.A.D.A. laboratory. It is claimed that this is stronger and more durable than wood wool cement boards.

Sawdust can also be mixed with cement to make a light weight concrete that has better insulating properties than the ordinary variety but its durability under exposed conditions is somewhat uncertain.

Within the scope of this chapter it has only been possible to indicate the wide range of products that can be made from timber. Many of these have superior strength qualities compared to the wood from which they have been made, and at the same time they have improved stability and increased durability. From an economic point of view they afford special opportunities for integrated forest industries as they permit a much higher proportion of the total output of the forest to be used, and by utilising much smaller trees shorter rotations of forest crops can yield a greater return to the growers. For all these reasons it seems certain that in the future more and more wood will be turned into panel products and other modified forms of timber.

BIBLIOGRAPHY

Akers, L. E. *Particle Board and Hardboard.* (Oxford, Pergamon), 1966.

Andrews, H. J. *An Introduction to Timber Engineering.* (Oxford, Pergamon), 1971.

Proc. Conf. on Development in the Use of Wood-Based Sheet Materials, *Jornal of the Institute of Wood Science* 5, (27), 1963.

Mitlin, L. (editor). *Particle Board Manufacture and Application.* (Dorchester, Pressmedia), 1968.

Sunley, J. G. *Finger Joints in Timber.* (London, British Woodwork Manufacturers Association), 1969.

T.R.A.D.A. *Plywood: Manufacture and Uses* Metric edition. (High Wycombe, T.R.A.D.A.), 1972.

Wood, 7th Annual Board supplement, April 1970.

CHAPTER THIRTEEN

Timber Bending

When man first learnt to twist small branches and twigs to make baskets the art of wood bending had begun. Other early examples of the craft are the bending of boards to make boats and of staves to make casks. Later bentwood furniture was manufactured and large quantities were imported into the U.K. from Central Europe. Most of these crafts are still practised today, though some only to a limited extent.

At the Forest Products Research Laboratory a comprehensive study of the factors influencing the bending of timber led to important technical developments. These have been described in a well illustrated handbook by Stevens and Turner. Here there is space for only a brief glance at the subject.

Bending of Solid Timber

When a piece of timber of square section is succesfully bent and both ends remain square it follows that the lengths of the concave and convex faces can no longer be equal. As wood cannot be stretched to any appreciable degree the difference must be the result of the shortening of the concave face. This is brought about by the compression of the fibres on the concave side which is the result of the pressure applied to the ends during the bending process.

33. Cooper driving iron hoop over steamed oak staves to form beer barrel. (W. Nurnberg)

Wood is a very elastic material and when the force required to bend a piece is removed, the wood will spring back to its original shape unless forcibly restrained. Some species of wood when heated in a damp condition become semi-plastic and their compressibility is very much increased, so that if the wood is heated throughout, before it is bent, there is much less risk of compression failure on the concave side of the end. Heating to 100°C is most conveniently done in a steam box. About three-quarters of an hour per inch thickness (1·8 minutes per mm) is generally required to heat wood throughout.

Bending Methods

The devices for bending wood by hand into simple shapes and the more elaborate machines for making larger and more complicated bends have been described in detail by Stevens and Turner. Many of these devices involve the use of a steel strap to support the outer face in order to prevent tension failures of the outer fibres. This is combined with an end-stop which can be adjusted so as to exert a compressive pressure on the concave face. After the wood has been bent the ends are clamped or tied together until it has set in its bent shape. This process of setting is much accelerated by exposing the bent wood, while it is still under constraint, to hot dry air, but even so it requires many hours. As an example, Stevens and Turner suggest that nine hours at 66°C are required to set a 32 × 32 mm piece bent to a radius of 230 mm. This is then allowed to cool for an hour and then stored under workshop conditions for two days before use.

Choice of Timber for Bending

In general softwoods cannot be bent so easily as can certain tough hardwoods, nor can many of the tropical hardwoods be bent easily. The bendability of a very wide range of timbers has been investigated at the Princes Risborough Laboratory. This quality can conveniently be expressed as the limiting minimum radius of curvature of one-inch sticks of air-dried and steamed material.

The tough hardwoods of Europe and North America that possess outstandingly good bending properties include ash, beech, elm, oak, robinia, sycamore and walnut.

Wood for bending must be straight-grained and free from knots and from the slightest trace of decay. It should have a moisture content of 25–30 per cent when it is bent. Freshly felled green timber is liable to rupture when bent as a result of hydraulic pressure in the cells which are full of water.

Laminated Bending

It is well known that very thin strips of wood can easily be bent to a small radius of curvature without breaking. It is therefore easier to build up a thick curved piece by gluing together a number of thin laminations than to attempt to bend a large piece in which severe stresses must occur. With the advance of modern waterproof adhesives that can produce bonds as strong as the wood itself many articles (such as tennis rackets for instance) that formerly were made by bending solid pieces of wood, are now manufactured by laminated bending with laminations about 3·2 mm ($\frac{1}{8}$ in) thick. It is possible in this way to produce satisfactory bent articles from timbers that cannot be bent in solid form, and also from timber of poorer quality. The laminae should always be dried before gluing to a moisture content not exceeding 20 per cent and preferably to one that will be in equilibrium with that likely to prevail in the conditions of use.

Laminated bends are particularly useful when bends of small radius are required, (as for hockey sticks) and where there is always a rather high proportion of failures when they are attempted from the solid wood.

As an alternative to gluing together bent laminae, ready manufactured plywood can itself be bent. Curved shapes can be made by taking flat sheets of plywood and bending them as required around suitable forms or moulds. These have been made to produce parts of chairs, radio and television cabinets, and containers of various types. Stevens and Turner have described a number of devices for making rounded corners in thick plywood. Plywood bonded with WBP or BR glue should be steamed or dipped in very hot water for a few minutes to soften it before bending. That bonded with an MR adhesive should be soaked in water at a temperature not exceeding 67°C.

The use of curved laminated structural members is referred to in Chapter 12. The technical problems involved in their manufacture are summarised in *F.P.R.L. Technical Note* No 17. In this it is pointed out that if boards are shorter than the member being produced they can be end-jointed by finger joints. After jointing, the laminations must be planed to a uniform thickness which is usually between 32 and 44 mm for straight pieces; for curved members, thinner laminations (13–25 mm), that can easily be bent without fracturing, are preferable.

After glue has been spread on both surfaces of inside laminations and on the inner surfaces of the outside laminations they are assembled in a jig and a bonding pressure is applied. For softwoods this must be at least 0·7 N/mm² and for hardwoods 1·0 N/mm². The pressure in the clamps is maintained for 6–8 hours if the temperature of the workshop is not less than 10°C.

After removal from the jig the glue squeezed out is removed by scraping, but if the member is intended to be visible the surfaces should be planed. They are then ready to be varnished.

BIBLIOGRAPHY

Stevens, W. C. and Turner, N. *Wood Bending Handbook.* (London, H.M.S.O.), 1970.

CHAPTER FOURTEEN

Defects in Wood

The influence on the strength of wood of certain growth features such as knots, deviation in the slope of the grain, and compression failures, is described in Chapter 4. In this chapter defects that develop in felled timber are considered. Many are the result of faulty seasoning procedures or of delay in extracting and converting the logs.

Seasoning Degrade

Wood shrinks as it dries and this shrinkage is not uniform in all directions. The exposed surfaces tend, of course, to dry out more rapidly than the interior. For these reasons tensions are set up in the wood as it dries down below the fibre saturation point, and this may result in permanent distortion or warping, or to splitting along the lines of weakness, that is, along the rays.

After a tree is felled the cut end is exposed to the air and if the weather is hot and dry rapid drying out may result in deep radial splits developing. This damage may be greatly mitigated by applying a moisture-resistant coating to the end of the log immediately after the tree has been felled. Sometimes spiked metal braces, known as 'dogs' are driven into the ends of logs or large baulks of timber in an endeavour to check this end splitting. However, as explained on page 87, the best way to protect logs of susceptible species awaiting conversion is to submerge them in a log pond or to keep them soaking wet by spraying them continuously.

Woods vary greatly in their susceptibility to degrade during seasoning. Logs of light softwoods seldom split seriously and after conversion into boards they can be dried fairly rapidly, but they *are* susceptible to staining (see p. 192). Dense hardwoods, especially those with broad rays such as oak and beech, can suffer severe degrade if exposed to unduly rapid drying and for this reason it may be preferable when possible to fell such trees in the cooler, wetter seasons of the year. Some of the very dense woods, such as box, require special protection during the first stages of seasoning. It is usual to quarter the small logs of box and to leave them stacked, not too openly, in the shade for several years before converting them into boards or strips.

The severe splits that appear in the ends of logs, or in large baulks of timber, are called '*shakes*'. If these originate at the pith and spread out along the rays they may form what is known as a '*star shake*' while a crack developing tangentially along a growth ring may cause a '*ring shake*' separating a central core from an outer ring. When damage of this kind occurs it is impossible to saw intact boards from the affected ends of the logs and much useful timber is wasted.

Sawn timber is similarly liable to degrade during seasoning if this is carried out improperly or if the drying conditions are unduly severe. Since the shrinkage in the tangential direction is greater than in the radial, the side away from the pith tends to shrink more than the opposite face. If the board is not restrained during drying permanent warping across the width results in a condition known as '*cupping*'. If the grain of the wood is interlocked or spirally distorted permanent '*twisting*' of the board may occur.

Both cupping and twisting can be greatly reduced if the piles of wood are built high and weights are applied to the top of the pile to keep the boards flat while they are drying out. The piles should also be covered to protect them from sun and rain. These defects are seen most frequently in the upper layers of seasoning piles where this precaution has not been taken.

Boards from the centre of the tree sometimes bend in the longitudinal plane without cupping. This form of distortion is called '*spring*' and is due to the release of stresses that existed in the tree.

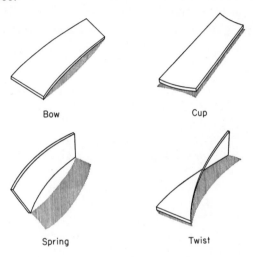

Bow   Cup

Spring   Twist

34. Distortion of boards on drying.

If the surface layers of planks are dried too quickly while the core remains wet, severe surface tensions are set up and the timber may become what is termed '*case hardened*'. Then, as the core dries out it shrinks inwards from the 'set' outer layer, resulting in internal '*honeycomb*' checks appearing. This condition can be remedied by steaming the wood in a kiln so as to bring the whole piece into an equilibrium moisture content.

Some timbers suffer '*collapse*' during kiln drying. This generally occurs in rather weak timbers in which the vessels, or tracheids, cave in when the wood dries. It occurs most frequently in fast grown immature trees (e.g. some home grown Sitka spruce) and in the more porous early wood. It leads to corrugation of the surface, a condition known as '*wash boarding*'. Collapse can largely be removed by steaming the dried timber in a saturated atmosphere for a few hours.

Staining

Discolouration and staining of pale-coloured timbers may develop in felled logs or in sawn timber during seasoning. This can be attributed to one of the following: sap-stain (bluestain) due to moulds; incipient decay (dote); contact with chemical, particularly iron; overheating leading to charring; oxidation of cell contents; or weathering.

Sap-stain

Sap-stain is a greyish-blue discolouration that develops in the sapwood of pines and is caused by the growth through the wood of moulds having dark-coloured hyphae. A similar discolouration can occur in poplar and also in many of the light-coloured tropical hardwoods such as obeche and ramin. This condition does not significantly weaken the wood and must not be regarded as an indication of incipient decay (dote). It does however spoil the appearance of light-coloured woods that are to be used with a clear finish, or which particularly need to present a clean look, for example in packing cases for foodstuffs. The price paid for stained timber may, therefore, be appreciably less than that of the same grade of timber that is free from stain.

To prevent staining of logs in warm weather it is necessary to get them out of the forest very quickly. Under tropical conditions a delay of only a few days may allow deterioration by staining fungi and insects to become established. If logs cannot be sawn up soon after felling, preservative action should be taken as outlined in Chapter 7.

Effective fungicides are now available which can be applied to the freshly sawn timber and that will protect it until the wood has dried down below the moisture content at which staining moulds and fungi can develop. Only dilute solutions averaging about 1·0 per cent (i.e. 1 kg of solid in 100 litres of water) are required to protect the wood if it is going to be piled openly at once. But rather higher concentrations (up to 3 per cent) may be needed to protect green timber that is to be shipped in solid packages. The most effective chemical that has so far been discovered for preventing sap-stain is sodium pentachlorophenate. Formulations of this type are sold under various trade names such as Brunobrite and Santobrite. Often a certain proportion of borax is added. This not only makes a cheaper solution but it improves the solubility of the fungicide. Sodium pentachlorophenate is poisonous, but if the maker's recommendations are followed carefully the dilute solutions can be used without risk to health. Operatives handling timber wetted with the fungicide should always wear rubber gloves to prevent the possibility of the solution causing irritation to the skin.

Where kiln drying facilities are available staining can be prevented by drying the timber in a kiln immediately after it has been sawn from the log. It is, however, often more economical to air dry the timber to a certain extent, before putting it in the kiln for the final drying down to the moisture content required. In this case fungicidal treatment may be necessary to keep the timber clean during the air drying period.

Under humid tropical conditions it is often impossible to season certain timbers that are particularly susceptible to sap-stain, such as obeche and ramin, in the open without deterioration. Treatment with a fungicide is therefore essential if they are to be shipped in a clean condition.

Total immersion of the sawn timber in a solution for a minute or so is sufficient to give protection against sap-stain. If only small quantities have to be treated this can be done by hand, provided the operatives are suitably protected against any solution reaching the eyes or skin. However in any mill where there is a large volume of timber to be treated it is worth while to install a dipping tank provided with a mechanical device for passing the timber through the solution on to a draining board from which excess fluid can drip back into the tank and so avoid waste.

Staining due to Incipient Decay

Changes in colour due to incipient decay involve the development of dark brown patches or streaks. In the case of hardwoods bleaching in patches may occur, or narrow dark lines. The fibres will be found to break off short when tested with a knife as incipient decay quickly lowers the toughness of the wood. Timber affected in this way must be rejected for any purpose in which strength is important. It is admissable only in the lowest grades of timber to be used for rough purposes. Prevention of decay has been discussed in Chapter 7.

Staining Due to Chemicals

Contact with iron is a very common cause of a bluish-black stain in woods such as oak and sweet chestnut that contain large amounts of tannin. The tannin reacts to form iron tannate, which is the basis of ordinary ink. This kind of stain often appears as dark streaks on oak fencing when ordinary nails have been used to fix the palings. It can be prevented by using galvanised nails and fastenings. Iron stains do not usually penetrate very deeply and they can be removed by sponging the surface of the wood with 6 per cent oxalic acid. This should be washed off after removal of the stain.

Alkaline solutions such as ammonia cause darkening in many hardwoods. Exposure to ammonia vapour is the means by which the dark-coloured 'fumed oak' is prepared.

Staining Due to Overheating

It is usually fairly obvious when a dark brown stain has been caused by scorching. This can happen when wood has been cut by a saw that has been binding, or when a piece has been impeded in its passage through a planer.

Weathering

The colour of wood that is fully exposed to the weather in this country generally fades, and in clean air the surface becomes a pleasing silvery-grey colour. This is due to the washing out by rain of the coloured constituents of the wood. In very dry climates, however, wood long exposed to sunlight with intensive ultra-violet radiation becomes darkened.

Ambrosia (Pinhole Borer) Beetle Damage

As described on page 93 pinhole borers cause degrade in logs of many tropical hardwoods such as afara, obeche, abura, agba and meranti, and also in certain softwoods such as Douglas fir and parana pine.

It has been found that thorough spraying of all surfaces of a log with BHC-miscible oil concentrate diluted with water to give 0·4 to 0·75 per cent gamma BHC affords a high degree of protection. Various proprietary preparations of this chemical have been formulated for protecting logs against pinhole borers.

Only in the lowest grades of American and Far-Eastern hardwoods are small borer holes permitted and so their presence seriously reduces the value of the timber in which they occur. Therefore from a commercial point of view it is most important to prevent or minimise the attack of susceptible species by these insects.

BIBLIOGRAPHY

Bateson, R. G. *Timber Drying and the Behaviour of Seasoned Timber in Use* 3rd edition. (London, Crosby Lockwood), 1952.

Bletchley, J. D. *Insect and Marine Borer Damage to Timber and Woodwork*. Forest Products Research Laboratory (London, H.M.S.O.), 1967.

Ertfeld, W. *et al. Defects in Wood*. (London, International Textbook), 1964.

Findlay, W. P. K. *Timber Pests and Diseases*. (Oxford, Pergamon), 1967.

CHAPTER FIFTEEN
Grading of Timber

The natural variability of timber is so great that exporters soon found it necessary to grade the product from their mills so that buyers should know the quality of the goods they would receive. Attempts have been made to standardise the grades nationally and associations of timber producers in Canada and the U.S.A. have laid down precise rules for grading the timber of the areas in which they operate, as have various other regions around the world.

The basis on which softwoods and hardwoods are graded are so different that they must be considered separately.

Grading of Softwoods

The grading of softwoods may be based either on appearance or strength.

Appearance—the absence or presence of visible defects. The main defects taken into consideration include knots, shakes, splits, wane, twists, spring, cupping, discolouration and decay. This was the traditional way in which sawn timber was graded.

Strength. By this method an attempt is made to ensure that every piece included in a particular grade has a certain minimum strength. (see p. 202).

In the United Kingdom five grades of sawn softwood are recognised in BS 3819. Grades I Clear and I are generally suitable for joinery and high-class structural purposes. Grade II is suitable for structural work and carcassing, while Grades III and IV can only be used for general purposes where consistent strength properties are not of prime importance.

The bulk of the softwood produced in northern Europe, Poland and eastern Canada and exported to this country is now graded ('bracked') in two or three main classes:

Unsorted—the top quality which includes the timber that was formerly graded into classes I to IV—except in the U.S.S.R. where Unsorted includes I to III; *Quality V* (= U.S.S.R. IV); and *Quality VI* or 'Utskott'.

It is necessary to have some knowledge of the shippers and of the ports from which the timber has been exported in order to appreciate the relative qualities of the redwood shipped from Russia and the Baltic. Broadly speaking the further north the timber has been grown the finer will be its quality owing to the slower rate of growth of the trees. The finest joinery wood from the U.S.S.R. is obtained from trees grown in Siberia and shipped from the Kara Sea. The best Swedish redwood comes from the forests in the north and is shipped from ports in the Gulf of Bothnia. The best quality whitewood, however, comes from the lower part of the Gulf of Bothnia and from the White Sea.

The origin of any parcel of softwood from the Baltic can be identified by examining the shipping mark which is usually stencilled in red on the ends of the pieces. The initials of the exporter, which are sometimes combined with a crown or other symbol, are used to indicate the source and grade of the wood. Timber from the U.S.S.R. is hammer-stamped with a combination of letters to indicate production, grade and port of shipment. For details of shipping marks reference should be made to a special publication listing the majority of them (see Bibliography).

Timber shipped from the Pacific coast of North America is

inspected by the Pacific Lumber Inspection Bureau and is graded into the following qualities:

Nos. 2 & 3. Clear and Better. This comes from the flanks of large logs and must not contain more than a few small defects and these only on one face. It is suitable for high class joinery.

Selected Merchantable. This is sound, strong, close-grained timber suitable for good constructional work. Small sound knots are permitted.

Nos. 1 & 2 Merchantable. These grades permit larger defects but provide sound structural timbers for carcassing.

No. 3 Merchantable. This lowest grade permits all kinds of defects.

There are separate grading systems for specific end-purposes such as doorstock, flooring, mining timbers, and so on.

In Brazil four grades are recognised but only Grades 1 and 2 are exported to Europe.

The comparison between these various general grades set out in the Table below can only be approximate.

Grading of Hardwoods

The rules for grading hardwoods that have been adopted in some of the major exporting countries, and more recently in the U.K., are based on a method known as the 'cutting system'. By this method an estimation is made of the amount of material free from all defects (or having only acceptable defects) that can be obtained from a plank. This is assessed in terms of rectangles and measured in units of 12 in^2 (77·4 cm^2). Each grade specifies the minimum area of cuttings acceptable and expresses this as a fraction of the total area. The number of cuts allowed in a grade and the minimum size of each cutting depends on the size of the plank.

When hardwood logs are converted directly into large pieces for special purposes, such as railway sleepers for instance, the products are graded on the number and size of the defects, as in the grading of softwoods. The sawmillers dealing in home-grown timbers generally grade in this way as

TABLE 6. Approximate comparison of softwood grading systems

UK BS 3819	I Clear	I, II	III	IV
Norway, Sweden, Finland & Eastern Canada		I, II, III, IV Unsorted.	V	VI
USSR		I, II, III Unsorted.	IV	V
Brazil	No. 1 & 2			
British Columbia and Pacific coast of North America (R. List)	No. 1 Clear No. 2 Clear No. 3 Clear	Select Merchant-able No. 1 Merchant-able	No. 2 Merchant-able	No. 3 Common

Reproduced from B.R.E. Princes Risborough Laboratory *Technical note* No. 27 CCR.

TABLE 6a. Preferred sizes in stress graded timber

	75 mm	100 mm	125 mm	150 mm	175 mm	200 mm	225 mm
38 mm	×	×	×	×	×	×	×
50 mm	×	×	×	×	×	×	×
63 mm				×	×	×	×

These sizes can be specified in visual and machine-graded GS and SS, and machine-graded only in the highest grade, M75, all to BS 4978. While the whole range of sizes listed in BS 4978 will be stress-graded, if specifiers work to the 18 preferred sizes they should have no difficulty in obtaining material.

most of the hardwood from their mills is sold for a specific purpose.

Below is reproduced a comparison of the grading rules of interest to importers and users of hardwood in the U.K; but it must be noted that considerable quantities of sawn hardwood are also imported from other regions in which no precise grading rules apply.

TABLE 7. Summary of features included in existing hardwood grading rules

D—Considered a defect
DX—Considered a defect if excessive
X—Acceptable defect
BLANK—Defect not considered

RULES	Characteristics and Defects	Cuttings (Fraction Clear)	Lyctus	Brittle Heart	Borer Holes Small/Med	Borer Holes Large	Decay (Rot)	Knots Pin	Knots Sound	Knots Un-sound	Mineral Streaks and Spots
N.H.L.A. (U.S.A.)	1st	11/12			DX	D	D	DX	DX	DX	DX
	2nd	10/12†			DX	D	D		DX	DX	DX
	Selects	11/12			DX	D	D	DX	DX	DX	DX
	No. 1 Common	8/12†			DX	D	D	DX	DX	DX	DX
	No. 2 Common	6/12†			DX	D	D	DX	DX	DX	DX
	No. 3a Common	4/12			DX	D	D	DX	DX	DX	DX
	No. 3b Common	3/12			X	D	D	X	X	DX	X
	Below Grade				X	X	D	X	X	X	X
ASIA PACIFIC	Prime	10/12†	D	D	DX	D	D	DX	DX	D	
	Select	9/12	D	D	DX	D	D	DX	DX	D	
	Standard	8/12	D	D	DX	D	D	DX	DX	D	
	Serviceable	8/12	D	X	X			D	X	X	D	
BS 4047 (U.K.) CUTTING	1	10/12 (11/12 Unedged)	D				D	DX	DX	D	
	2	8/12 (10/12 Unedged)	D				D	DX	DX	D	
	3	6/12 (9/12 Unedged)	D				DX	X	X	D	
	4	3/12 (8/12 Unedged)	X				DX	X	D	D	
BS 4047 (U.K.) DEFECTS	A		D		D		D	DX			
	B		D		D		D	DX			
	C		D		DX		DX	DX			
	D		D		X		DX	DX			

† Fraction increases with decreasing size of piece.
* Acceptable if bright.
Reproduced from F.P.R.L. *Technical Note* No. 33. C.C.R.

Pith	Resin or Bark Pockets	Size Variations Width	Length	Slope of Grain	Stain	Sapwood	Splits	Shakes	Seasoning Checks	Wane	Warp
		90% full									
DX		6 inch†	8–16 ft		DX	X	DX	D	X	DX	DX
DX		6 inch†	8–16 ft		DX	X	DX	D	X	DX	DX
DX		4 inch†	6–16 ft		DX	X	DX	D	X	DX	DX
DX		3 inch†	4–16 ft		DX	X		D	X	DX	
DX		3 inch†	4–16 ft		DX	X		D	X	DX	
DX		3 inch†	4–16 ft		DX	X		D	X	DX	
DX		3 inch†	4–16 ft		X	X		D	X	DX	
X					X	X		X	X	X	
D	DX	6 inch†	10 ft†		D	D	DX	D		DX	DX
D	DX	5 inch†	8 ft†		D	DX	DX	D		DX	DX
D	DX	4 inch†	6 ft†		D	X*	DX	D		DX	DX
D	X	4 inch†	6 ft†		X	X If Sound	DX	D		X	DX
D (X Sq Edge)		6 inch†	6 ft†	DX	D	X*	DX		DX	D	DX
D (X Sq Edge)		6 inch†	6 ft†	DX	D	X*	DX		DX	DX	DX
X		6 inch† (4 inch Sq Edge)	6 ft†	DX	DX	X*	DX		DX	DX	DX
X		6 inch† (3 inch Sq Edge)	6 ft†	X	DX	X*	X DX Sq Edge		DX	DX	DX
D				DX	D	X*	D		DX	D	DX
DX				DX	DX	X*	D		DX	DX	DX
X				DX	DX	X*	DX		DX	DX	DX
X				DX	DX	X*	DX		DX	DX	DX

Stress Grading of Timber

Traditional visual grading rules depend on the evaluation of defects present in the sawn pieces, but the significance of these defects depends on the purpose to which the timber is to be put. For instance growth features, such as sloping grain, which do not mar the appearance of the wood, may have a very serious effect on the strength and render it unfit for such purposes as ladder stiles. Conversely blemishes such as sap-stain that spoil the decorative appearance of the wood may have no appreciable effect on its strength, and the presence of these may be quite acceptable in carcassing timber.

Stress (or strength) grading is the classification of the sawn pieces of timber intended for structural purposes so that every piece within the given grade has a certain minimum strength value, or, more precisely, that the strength of all the pieces classed in that grade falls within a certain range.

Unless an architect or engineer can know with some degree of precision the mechanical properties of the materials he is going to use he must use larger sections to make allowance for the possibility that some pieces may fall much below the average of the rest. The strength property of a manufactured material such as steel can be predicted quite accurately, but the strength of timber varies considerably from piece to piece even though these may come from the same species or even have been cut from the same tree.

Stress grading by observation of the distribution and size of knots has been practised succesfully for some time past in America, but European saw millers have been unwilling to change their long established grading rules. The extra price that could be charged for stress graded material was not sufficient inducement to warrant the cost of the labour involved in the operation. Although some of the growth features, such as knots, can be assessed quite quickly by a trained operative, others such as density and slope of grain involve a time consuming scratch test and cannot be assessed by eye.

Extensive studies at Princes Risborough and elsewhere have shown that by measuring the stiffness of a board its ultimate bending strength can be assessed with a fair degree of accuracy. By making use of this relationship a method for testing every board mechanically under a light, nondestructive, load has been devised. The Computermatic can automatically sort at high speed pieces of timber into strength grades more reliably than any visual inspection. The cost of this machine was £14 000 in 1973, and it has been estimated that for an output of 25 000 m³ per year the total cost of stress grading by machine was £1·05/m³. It has also been estimated that the labour cost of visual grading was 13 p/m³, but machine grading can give a higher yield of the upper grades as the machine takes density into consideration.

The publication of the British Standard specification BS 4978 for timber grades for structural use will give a great impetus to the use of stress graded wood in the United Kingdom and dealers in Sweden are already preparing to supply it. The rules conform to similar ones used throughout North America. With the higher stress values that can be assigned to machine graded timber it becomes possible to justify using smaller sizes of joists for conventional floor constructions and house carcassing elements. Grade stresses for European redwood and whitewood were published by Curry and Covington in 1969. Limiting spans for machine-graded European redwood and whitewood have been published by the Princes Risborough Laboratory to enable advantage to be taken of the greater efficiency of machine stress grading.

BIBLIOGRAPHY

British Standards Institution. *Grading Rules for Sawn Home-Grown Softwoods.* BS 3819, 1964.
British Standards Institution. *Timber Grading.* BS 4978, 1973.
Curry, W. T. *Interim Stress Values for Machine Stress Graded European Redwood and Whitewood.* Timberlab Paper No. 21, 1969.
Forest Products Research Laboratory. *Sawn Hardwood Grading Systems.* Technical Note No. 33, 1969.
Hearmon, R. F. S. and Rixon, B. E. *Recommended Limiting Spans for Machine Stress Graded European Redwood and Whitewood.* Timberlab Paper No. 30, 1970.

CHAPTER SIXTEEN

Timber Trade of the United Kingdom

Great changes have taken place in the world trade in timbers during the present century and it is likely that further changes will occur in the future. One can only indicate, therefore, from whence supplies are coming to this country at the present time and where the forests are that contain the reserves of timber. Great changes in price have also occurred. The average price of softwood imported in 1914 was £2·18/m³ and in 1973 it was £37·00/m³.

Softwoods

Great Britain imports about 95 per cent of its timber requirements, but this figure is likely to drop as the supply of softwoods from the new plantations of the Forestry Commission increases. Traditionally the United Kingdom has imported the bulk of its utility softwoods from the Baltic and from Russia. Considerable quantities also came from Canada and these were increased owing to the stimulus to 'buy British' encouraged by the preferential tariffs agreed at the Ottawa Conference of 1932. During the Second World War all supplies of timber from the Baltic were cut off and for some time after the war Russia required most of her own production for rebuilding and reconstruction. Today however we have reverted to the old pattern, most of our softwood coming from the Baltic and Russia and only 10 per cent from Canada, half the volume received in 1964–7. Most of the Canadian production now goes to the U.S.A. and Japan.

It seems likely that we shall continue to import the bulk of our softwood supplies from the Baltic and Russia as in the well managed forests in these countries the increments are equal to, or in excess of, the removals. If the immense, hitherto untapped, reserves of softwood in the remoter parts of Siberia can be made available by improved communications there should be ample supplies of good quality softwood for many years to come. In Canada it is estimated that the annual increment considerably exceeds the annual cut so there should be no overall shortage there either.

Of our imports of sawn coniferous softwoods in 1972 the principal supplies (in round figures) were:

Sweden	2·4 million m³
Finland	1·9 million m³
U.S.S.R.	1·6 million m³
Canada	768 thousand m³
Poland	463 thousand m³
Portugal	173 thousand m³

Of softwood imports European redwood amounted to 2·5 million m³, whitewood (spruce, silver fir) 1·3 million m³ and mixed redwood and whitewood to 2·7 million m³.

Hardwoods

Prior to the First World War considerable volumes of hardwood were imported from the U.S.A. and Canada, but these countries now consume most of the American production and our supplies of sawn hardwoods come mainly from Africa and Malaysia.

In 1972 the following quantities (in round figures) of sawn hardwoods were imported from the major sources of supply:

	thousand m³
Malaysia	170
Ghana	108

Roumania	75
France	60
Ivory Coast	54
Denmark	48
Nigeria	22

Of sawn hardwoods beech represented nearly one third of the total import.

The bulk of the logs imported for peeling into veneers and for sawing came from:

	m³
Ivory Coast	98
Ghana	53
Nigeria	38

The principal species imported in log form were mahogany (African), utile, iroko, afrormosia and obeche.

Wood-Based Panel Products

The importation of wood in the form of plywood and other panel products is continually increasing as the use of these materials expands.

Finland has been our main source of supply for plywood for many years, but sizeable quantities are now also imported from Malaysia and West Africa. Our supply of Douglas fir plywood comes from Canada. In 1972 our imports of plywood amounted to 848 thousand m³ and of blockboard and laminated board to 213 thousand m³.

The imports (in round figures) from our principal suppliers came from:

	thousand m³
Finland	306
Canada	212
Malaysia	111
U.S.S.R.	87
Spain	32
Japan	30

Particle Boards

Again Finland is our main source for imported particle boards sending us 150 thousand tonnes in 1972. Norway and Sweden sent 55 and 53 thousand tonnes respectively.

Fibre Boards

Sweden is our main supplier of fibre boards sending 137 thousand tonnes in 1972 with Finland in second place with 53 thousand, and the Republic of South Africa third with 21 thousand tonnes.

Home production of fibre boards is expected to increase greatly in 1973 when three new plants, with a combined capacity of 180 thousand tonnes, come on stream. This will increase the proportion of home produced board from 28 per cent in 1972 to an estimated 38 per cent in 1975.

The Future

Well-informed authorities do not consider that there will be any world shortage of timber in the foreseeable future. Professor Richardson foresees a shift towards the tropical hardwoods, the trade in which has increased from 4 to 35 million m³ in 20 years; and though the export of logs from West Africa is likely to fall, the exports of sawn timber from the Far East and South America would make up for this. It is generally agreed that the grouping together of similar species that have the same end uses would facilitate the use of tropical hardwoods.

Latin America contains over 25 per cent of the world's forests with huge areas of natural forest in Columbia, Brazil, Peru and Venezuela. Much of this, however, is very difficult of access and it is not thought likely that any major proportion of our imports will come from these countries within the next ten to twenty years.

All authorities agree that the proportion of reconstituted composite wood products to solid timber will continue to increase. Particle boards, which can be made from relatively

small trees grown on a short rotation, will continue to replace
solid timber for many purposes where uniformity of material
is important. The following figures illustrate this trend very
clearly.

TABLE 8.   World consumption of forest products

Material	Unit millions	1955	Year 1965	1975 estimated
Sawn wood	m³	303	384	424
Plywood	m³	10·4	25·7	44·0
Fibre board	tonnes	3·0	5·6	10·9
Particle board	tonnes	0·5	5·2	16·5
Wood pulp	tonnes	46	74	130

The increased export of panel products from Malaysia from
2500 m³ in 1964 to about 98 000 m³ in 1972 is one example of
how production is increasing in the countries where the trees
are growing.

Integrated Wood Utilisation

Wood is one of the few renewable raw materials in the world
and present day concern about future resources of fuel, and of
raw materials with which to make plastics, is focussing attention on the use of wood as a source of chemical raw material
as well as of fibres. Egon Glesinger was an enthusiastic ex-
ponent of the need to use the whole of the tree. In 1950 he
pointed out that the yield of sawn wood often represents only
50 per cent of the weight of the original tree, the rest being rejected as branchwood, slabs, off-cuts and sawdust. All this he
considered could well be turned into plastics, wood sugar,
alcohol and lignin fuel. At that time he was considered as a
visionary but today his views, with some modifications, are
widely accepted though they have not yet been put fully into
practice. Pulp mills, however, are a step in this direction and
can use up to 70 per cent of the trees processed by them.
    In the Centenary issue of the *Timber Trades Journal* (1973)

several contributors give their ideas as to the future uses of wood. They foresee in the coming century a steady diminution in the amount of sawn timber and a great increase in the processing of wood for its fibres and chemical constituents— though a certain amount of decorative timber will always be required for craft and reproduction work.

Many of the qualities of slow-grown hardwoods, such as durability, can be conferred on other faster-grown (and therefore cheaper) timbers by treatment with preservatives, while strength properties can be modified by lamination and by the skilful marriage of timber with other materials.

So far as decorative hardwoods are concerned it is impossible to foretell how fashion will affect the demand for individual species or types; but it seems likely that there will always be a demand for the beautiful woods that delighted our ancestors, and probably for others as yet unknown in this country that may come to us from the forests of the Amazon Basin.

In the developed countries of Western Europe trees, in addition to being a source of timber, have an important role as attractive features of the landscape, and this amenity value may well in some areas outweigh the purely economic wish to obtain the maximum tonnage of raw material from every acre of forest. In fact the Government has already made a statement to the effect that whereas in earlier years the primary responsibility of the Forestry Commission had been to build up a strategic reserve of softwood, forestry should in future be part of a really effective pattern of rural land use harmonised to the best possible advantage with agriculture and the environment, together with opportunities for recreation.

BIBLIOGRAPHY

Glesinger, E. *The Coming Age of Wood*. (London), 1950.
The Changing Pattern of Forest Resources. *Timber Trades Journal*. November 18, 1972.
Leigh, J. H. *The Timber Trade—an Introduction to Commercial Aspects*. (Oxford), 1971.
Timber Trades Journal. Centenary Issue, 1973.

CHAPTER SEVENTEEN

Wood and the Arts

Wood being a material that is relatively easy to work with quite simple tools and one that is readily available in most parts of the world, it has been used since a very early date by sculptors, artists and musicians.

Woodcarving

In almost every civilization and since very early times the carving of wood has been practised, from simple peasant work fashioned with a knife by the fireside in the winter evenings, to the elaborate and ornate work of craftsmen decorating Baroque churches and cathedrals.

Almost every kind of wood has been used at one time or another by wood carvers, but when there is a choice they generally prefer a wood that has a close even texture and does not splinter; one that can also be seasoned without much distortion and which is relatively stable in changing conditions of humidity.

Among softwoods Western white pine and yellow pine are ideal for carving as they have a fine even texture and are very stable. Of the hardwoods the timbers of fruit trees, such as pear, apple and cherry, have long been popular. Lime and sycamore can be carved to give very fine results as can be seen in the work of Grinling Gibbons (1648–1720) which was mostly executed in lime wood; but walnut, mahogany and teak have more interesting figure and colours. Very hard timbers, such as box, ebony and rosewood, require patient working after the wood has been fully seasoned down to the moisture content it will attain in its final situation.

For larger carvings oak has been used extensively in England and splendid examples of fifteenth and sixteenth century wood carving can be seen in many screens, font covers and misericords in parish churches and cathedrals all over the country.

Wood carving has been done in almost every country of the world. Some of the most beautiful and delicate has come from China and Japan. Many of these pieces were covered with gesso (fine plaster mixed with glue) which was richly coloured and gilded. Some of the mediaeval English tomb effigies were similarly covered with gesso and painted. Particularly lively wood carvings were made in West Africa. These were always cut from a single block of wood with an adze or knife. Probably some of the largest wooden carvings ever made were the totem poles of the American Indians.

The sculptor in wood often allows the natural growth and grain to influence the form into which he ultimately shapes his work, the lines formed by the annual growth rings thus enhancing its curves and design.

Wood for sculpting must be seasoned throughout down to an even moisture content. If the moisture content of the centre of a piece is appreciably different from the surface layers, splitting and distortion may ruin the work when the wood eventually dries out. If very large pieces of slow-drying wood, such as oak, are required, the only way to obtain these properly dried without years of delay may be to saw the piece into one inch boards, carefully dry these out, and then glue them together with a colourless glue. The large re-formed piece than has a uniform moisture content throughout its entire thickness.

The carver's principal tools are various sizes of chisel, and a mallet to drive these when the wood is very hard. Gouges, rasps and surform tools are also used.

Wood Engraving and Woodcuts

The earliest form of printed picture was made from blocks of wood on which a cut-out design stood up in relief. This was inked and pressed on to the paper. In the first printed books the text as well as the illustrations were cut out in this way.

The tool for making woodcuts is a sharp knife which is used to outline the shapes required and to cut away below the surface parts not intended to print. Large pieces are removed with a gouge. The wood used is usually pear or cherry and it is cut into slices along the grain of the trunk or branch. Wood engraving, on the other hand, is done with a triangular-shaped tool called a burin and a much harder wood is used, generally boxwood, which is cut across the grain so that the print is cut on the end grain. Much finer lines can be achieved in this process than in woodcutting.

Some of the finest woodcuts were made in the sixteenth century by Durer and Hans Holbein. Later the woodcut declined in popularity and by the end of the seventeenth century metal engravings had largely replaced woodcuts for illustrating books.

In the eighteenth and nineteenth centuries coloured woodcuts were produced in Japan, in large numbers and of great beauty and charm, and these were much appreciated in the West.

At the end of the eighteenth century the English engraver Thomas Bewick made a number of wood engravings of country scenes with studies of animals and birds in very fine detail. During the following century, however, wood engraving became merely a cheap method for reproducing drawings, until William Morris led a movement to encourage artists to design and cut their own and wood engravings of high artistic quality were again produced. Then in the 1920s several artists, notably Clare Leighton and Gwen Raverat, produced woodcuts as illustrations for books which have been greatly admired and which set a new standard in the art of woodcutting.

Wood as a Base for Painting

The earliest paintings were usually applied as decoration to walls—as in Pompeii—and pictures that could be carried from one place to another were a later development. Many of the greatest paintings of the Renaissance in Italy took the form of frescoes, but other than these the early Italian paintings were always done on wood. This was sometimes covered with canvas over which gesso was applied before the pigments in a tempera medium were painted on. The first oil paintings were applied to wooden panels often as triptychs in churches. Identification of the timbers used has sometimes assisted in the assignment of a painting to its correct period. For instance, if a reputedly mediaeval painting was found to be on a timber from the New World it could safely be assumed to be a later copy.

Paintings on wood can become seriously damaged if they are exposed to alternating damp and dry atmospheres. The movement of the wood as it swells and shrinks places a strain on the film of paint which may lead to its cracking and peeling off. This is particularly likely to happen when the back of the panel is unpainted so that this picks up moisture more quickly than the front and in consequence swells more than the painted surface. The whole panel therefore flexes with changes in moisture content and this has, of course, disastrous effects on the paint film. Great damage has been caused in this way to some paintings that were brought from damp churches into dry, centrally heated buildings. In modern picture galleries the atmosphere is conditioned to maintain as closely as possible the same relative humidity throughout the year.

When the wooden backing to a painting has crumbled away through decay and woodworm it is possible nowadays with very skilful work to remove the remains of the wood and to transfer the paint film to a new piece of canvas attached to a frame of a light wood such as balsa.

Inlay Work and Marquetry

The inlaying of slices or veneers of richly coloured woods into plainer ones so as to form an attractive design has been practised since classical times. It was introduced from Persia into Venice in the fourteenth century whence it spread into western Europe. It reached its apogee in France in the reign of Louis XV (1710–1774) when extremely elaborate designs were used to decorate tables, escritoires, bureaus, etc.

Veneered marquetry consists of building up a composite veneer of several contrasting woods which is then glued over a plain surface. There is much less labour involved than with inlaying as no grooves need to be cut out and also a number of veneers can be cut out in one operation. The craft of marquetry veneering is said to have originated in France and to have been introduced into England by Dutch workers about 1680.

In this century artists have made attractive pictures built up with pieces of veneer chosen so that the grain contributes to the design of the picture. The curling grain of the growth rings, for instance, cut tangentially, may emphasise the sweep of distant hills or clouds.

Tunbridge ware was made in Tunbridge Wells during the seventeenth century and continued to be made there until the beginning of the present century. It was very popular in Victorian times for domestic bric-à-brac. Elaborate compositions were made up of thousands of long, very thin, square pieces of different coloured woods glued together. A number of copies of the same design could be made by sawing thin slices off the master block, through the whole of which the pattern persisted—like the name through sticks of peppermint rock. Over a hundred kinds of wood were used to give the variety of tones required. Holly was used for white and this could be stained grey with Tunbridge Wells spa water. Green was provided by oak, or other hardwood, stained by the fungus *Chlorosplenium aeruginosum*. The use of this fungus to stain wood a brilliant emerald green was afterwards patented (in 1911) by F. T. Brooks, later to become Professor of Botany at Cambridge.

Musical Instruments

Most early musical instruments, apart from metallic trumpets and those made from animal horns, were of wood. They fall into two groups, stringed instruments and wind instruments.

Stringed Instruments

The harp is an instrument of great antiquity and many different varieties were known in Ancient Egypt. The wood used for harps today is chiefly sycamore, but the soundboard is of pine along the centre of which is glued a strip of beech, or other hardwood, in which are inserted the pegs that hold the lower ends of the strings.

Much timber goes into the making of pianos. The early issues of the *Timber Trades Journal* specifically included pianoforte makers among the readers for whom the Journal was intended. The soundboard is commonly made of Canadian Sitka spruce, or spruce from the Carpathians (which is known in the trade as 'Roumanian pine'). The bridges that hold the pins are made of beech or rock maple. For the case various ornamental hardwoods have been used. Rosewood was fashionable at one time, and many fine cases have also been made from mahogany and walnut, while an ebonised finish has often been popular, especially for grand pianos. Many cases are now veneered over plywood, blockboard or a stable wood such as obeche.

The violin family, i.e. violin, violas, cello and double bass, are made almost entirely of wood, and consist of some seventy different parts. Maple and sycamore are used for the back, handle, bridge, neck, scroll and ribs. The belly, bar blocks, linings and sound post are of European spruce. Both the spruce and the maple should have an even, moderately wide and perfectly straight grain and be completely free from knots. Ebony is used for the tail piece, finger board, nuts, screws, button and pegs, which last may also be made from boxwood or rosewood. The only wood considered to be suitable for the bow is brazilwood—*Caesalpinia echinata*.

35. Bonding the sound bars to the inside of the 'table' of a modern lute. (C.I.B.A.—Geigy)

Despite all this intricate work, mass production of violins in Germany and France in the nineteenth century brought the price down to an incredibly low figure. In Grove's *Dictionary of Music* (Vol IV, 1898) it is stated that the cost of making a 'trade' violin by a famous maker at Mirecourt was estimated, including profit, at four shillings and sixpence each! Grove comments that such instruments, if carefully set up 'can be made to discourse very tolerable music'!

There has recently been a revival of interest in mediaeval and Renaissance stringed instruments, and many fine new specimens of these are now being made. The making of fine-quality lutes in Cambridge has been described in a Ciba-Geigy leaflet. This states that spruce or cedar must be used for the flat 'table' of the instrument but for the curved staves which form the back of the sound chamber timbers of contrasting colours can be used alternately (see Fig. 35), or figured sycamore or maple used alone also gives a very pleasing appearance.

Wind Instruments

Probably one of the earliest ways in which man made music was by blowing over a pipe or series of pipes, or by blowing down a hollow wooden tube with a hole in one side of it, thus making the simplest form of whistle. From this developed an instrument with a series of holes which could be closed to give different notes, and so the recorder and flute came into being. Flutes used to be made of cocus wood, but this now being unobtainable, rosewood or African blackwood (*Dalbergia melanoxylon*) are used instead. African blackwood is used also for making other wood-wind instruments such as clarinets and oboes, and also for bagpipes. These last used to be made of hornbeam, holly or apple wood before imported timbers became available. The bassoon is so large that heavy dense hardwoods are not suitable and they are generally made of sycamore.

Pipe organs contain much timber and present special problems as they are usually in churches that are often unheated and damp and the wood is therefore exposed to varying humidities and temperatures. The stability of the timber used is, under these conditions, of paramount importance. Details of the woods used for the constructional parts of organs are given in the survey by Pearson and Webster. The console of church organs is now generally made of oak.

There are of course a number of other wooden instruments; the xylophone, for instance, whose very name implies that it is made of wood. This has keys that have to stand hard wear and give a clear ringing tone and no timber has been found for this purpose to excell Honduras rosewood. Space however does not permit of going into this subject in further detail.

BIBLIOGRAPHY

Oughton, F. *The History and Practice of Wood Carving*. (London, Allman), 1969.

Pearson, F. G. O. and Webster, C. *Timbers used in the Musical Instrument Industry*. (London, H.M.S.O.), 1956.

Index